John Landen

The Residual Analysis

A New Branch of the Algebraic Art, of Very Extensive Use, Both in Pure Mathematics and Natural Philosophy

John Landen

The Residual Analysis
A New Branch of the Algebraic Art, of Very Extensive Use, Both in Pure Mathematics and Natural Philosophy

ISBN/EAN: 9783742823601

Manufactured in Europe, USA, Canada, Australia, Japa

Cover: Foto ©Thomas Meinert / pixelio.de

Manufactured and distributed by brebook publishing software (www.brebook.com)

John Landen

The Residual Analysis

THE
Refidual Analyfis;

A

NEW BRANCH

OF THE

ALGEBRAIC ART,

Of very extenfive USE, both in Pure Mathematics, and Natural Philofophy.

BOOK I.

By JOHN LANDEN.

LONDON,
Printed for the AUTHOR; and fold by L. HAWES, W. CLARKE, and R. COLLINS, at the *Red Lion* in *Pater-nofter Row*.
MDCCLXIV.

PREFACE.

HAVING, some time ago, hit upon a new and easy method of investigating the binomial theorem, by a process purely algebraical; I was led to consider whether the means which enabled me to investigate that theorem would not be of use in the investigation of other theorems; and I soon found, that a method of computation depending on such means might be applied in many enquiries. Whereupon, believing such method of computation might be acceptable to the Mathematical World, I, with a view of publishing it, endeavoured to improve it, and to dispose the several articles relating thereto in such order as might conduce to the easy attaining a knowledge thereof. The result of my endeavours is the following treatise; in which the elements and common branches of the analytic art are purposely omitted, upon the supposition that the Reader is previously acquainted therewith; my design being only to teach a particular branch of that art, with its application to Geometry and Natural Philosophy. Which particular branch I have called the Residual Analysis; *because, in all the enquiries wherein it is made use of, the chief means whereby we obtain the desired conclusions are such quantities, and algebraic expressions, as by Mathematicians are denominated* residuals.

The principles of the common Algebra and Geometry having been thought insufficient to enable the Analyst to pursue his speculations in certain branches of science; new principles, very different from those before made use of, have, through a supposed necessity, been introduced into Analytics. The Fluxionists, following Sir Isaac Newton,

introduce

introduce an imaginary motion, and recur to the generation of quantities by a supposed flowing, or continual increase, of their parts. Mr. LEIBNITZ *and his followers, to avoid the supposition of motion, consider quantities as composed of infinitesimal elements; and reject certain parts of the infinitely small increments of quantities as infinitely less than other parts. In the* Residual Analysis, *(admitting no principles but such as were anciently received in Algebra and Geometry,) we neither have recourse to infinitesimals, nor to the principles of motion; but consider magnitudes as already formed, without any regard to their genesis, except in particular cases, where the manner of their being generated may be the proper subject of enquiry: And, as this Analysis is not less (if not more) useful than the fluxionary, or differential, calculus, it will consequently appear, that the Analytic art, founded, and carried on, upon such principles as were anciently received therein, (without the aid of any foreign ones relating to an imaginary motion, or infinitesimals,) is far more extensive than Mathematicians have hitherto reckoned it.*

Quantities infinitely small, and quantities infinitely less than quantities infinitely little, being incomprehensible; and the rejecting certain quantities as infinitely less than other infinitely little quantities, being, except in approximations, a very unsatisfactory (if not erroneous) way of getting rid of such quantities: The Principles of the method of Infinitesimals are certainly liable to some just objections, which cannot be made to the Principles of Fluxions *; *such infinitely small quantities as are almost continually under consideration in the first mentioned method, being no way concerned in the fluxionary doctrine, when it is explained and applied in a proper manner. If any thing can be said by way of objection to the fluxionary method, it is, that the new principles on which it is founded, though accurate, are not the genuine principles of Analytics, for the improvement of which, those principles were borrowed from the doctrine of motion: And that, although such borrowed principles may enable us to give very concise solutions to certain problems, yet perhaps we must not expect to bring the Analytic art to its utmost perfection, otherwise*

* Dr. MATY, comparing the method of Infinitesimals with that of Fluxions, says, "Celle-ci a donc le merite d'une plus grande exactitude, mais cet avantage "est racheté par la necessité qu'elle impose d'avoir recours aux principes du "mouvement." JOURNAL BRITANNIQUE, *Mois de Fevrier* 1751.

than

than by proceeding upon its own proper principles. What weight there may be in such objection, I shall not take upon me to determine. Yet I must confess, that, how natural soever it may be, in the resolution of certain problems relating to geometrical magnitudes, to consider such magnitudes as generated by motion, it seems to me not natural to bring motion into consideration in resolving problems purely algebraical: Nor does it seem natural, in resolving problems concerning the motion of bodies, to superinduce imaginary motions, and therewith bring into consideration the velocity *of* time, *the* velocity *of a* velocity, *&c. nor yet does it appear more natural, in the resolution of other problems, to make use of the fluxionary method, when (as is most commonly the case in that doctrine) the fluxions introduced into the process can only in a figurative sense be said to be the velocities of increase of the quantities called their fluents; such figurative expression being not the natural language of Analytics, but frequently, instead of conveying clear and distinct ideas, is confusedly employed in treating of quantities as generated by motion, which in reality cannot be conceived to be so generated*. Therefore I am induced to think, that, not only in the resolution of problems purely algebraical, but likewise in Geometry and Natural Philosophy, when an analytical process is requisite, and what is called the common algebra is insufficient, the* Residual Analysis, *which is founded (as I conceive) on the genuine principles of Analytics, is, for the most part, more properly applicable than the Fluxionary Analysis, which is founded on new principles borrowed as above mentioned. But however, very far from being positive in this matter, I freely submit it to the Judgment of the Public.*

In comparing the methods of computation just now mentioned, it may be observed, that, where the direct method of Fluxions might be applied, we, in applying our Analysis, compute the value of the quotient of one residual divided by another; and, where the inverse method of Fluxions would be applicable, we, from a given residual divisor and a particular value of a certain quotient, assign the correspondent dividend, of a particular Residual form. Now, the particular value, which we have occasion to consider of such quotient, being always to unity, as the fluxion of the first

* Such quantities are weight, *density*, &c.

member

member of the dividend to the fluxion of the first member of the divisor; it therefore often happens, that, though we proceed upon different principles, part of our process is the same as if we had pursued the fluxionary method: So that, in fact, many of the articles in this treatise may be of the same use in the doctrine of Fluxions as in this Analysis; and probably some of those many articles may conduce to the improvement of that doctrine, if the Reader, having studied it, should chuse to adhere to it.

The SUBSCRIBERS to this WORK, whose Names are at present, come to Hand, are

ROBERT Austin, *Esq*;
Mr. Thomas Allen, *Teacher of the Mathematics, in Spalding*, 6 copies.

Edward Brown, *Esq*;
Charles Balguy, M. D.
Mr. Abraham Beharrell.
Mr. Abraham Baley.
Mr. Stephen Bee, *Steward, and Land-Surveyor*.
Thomas Barker, *Esq*;
The Rev. Stanhope Bruce, A. M.
Mr. Reg. Brathwaite, *of St. John's College, Cambridge*.
The Rev. Mr. Barr, A. M.
Mr. Charles Brent.
Mr. William Brailsford.
Daniel Bayley, *Gent*.
Captain James Barnes.
Mr. William Bevil, 2 copies.
Mr. Anthony Birks, *Master of the Free-School in Donington, Lincolnshire*.
Mr. William Bothamley, *Clockmaker in Spalding*.
Mr. Samuel Beeken.

The Right Hon. the Lord Carysfort, *one of the Lords of the Admiralty*.
The Right Hon. Sir John Cust, *Bart. Speaker of the House of Commons*.
Jonathan Cope, *Esq*;
John Calcraft, *Esq*;
Charles Causton, *Esq*;
Charles Cæsar, *Esq*;
James Collier, *Esq*;
The Rev. J. Colson, A. M. F. R. S. *late Lucasian Professor of Mathematics in the University of Cambridge*.
Robert Clarke, *Gent*.
The Rev. Mr. James Clarke.
Mr. Martin Cole.
Mr. William Caldecott, *Surgeon*.
Mr. William Clifton.
Mr. John Corbet.

Daniel Douglas, *Gent*.
Dominick Donnelly, *Gent*.
Mr. John Dickenson.

Mr. Richard Edwards.
Mr. Langley Edwards.

The Right Hon. the Countess Dowager Fitzwilliam.
Samuel Forster, *of Grantham, Gent*.
Mr. B. Fernor, *Professor of Mathematics, and Member of the Royal Academy of Sciences at Stockholm*.

The Rev. Mr. Gibson, *late Prebendary of Peterborough*.
The Rev. Mr. Kennet Gibson.

William Hanbury, *of Kelmarsh, Esq*;
Mr. William Hutchinson.
Mr. John Hyde.
Mr Henry Hetley, *of St. John's College, Cambridge*.
Mr. Edward Hare.

The Rev. John Image, A. M.
Mr. William Jepson.

Robert Kelham, *Gent*.

The Right Hon. the Earl of Ludlow.
The Rev. Dr. Robert Lamb, *Dean of Peterborough*.
Mr. William Lax.

Thomas Moore, jun. *Gent*.
The Rev. George Moore, A. M.
Mr. William Moore.
The Rev. Mr. Mirehouse.
The Rev. Mr. Marshall, A. M. *Fellow of Christ College, Cambridge*.
Mr. Francis Maseres, A. M.

Mr. Edward Nott, *Bookseller, and Teacher of the Mathematics, in Stamford*.

Walden Orme, *Esq*;

Armstead Parker, *Esq*;
Peterborough Society.
The Rev. Peter Peckard, A. M.
Mr. John Parish.
Mr. John Perry.
Mr. Joseph Priestley.

Mr. Thomas Quanborough.

Mr. Edward Rollinson

Mr. Richard Robinson, *Land-Surveyor.*

Dr. Robert Smith, F. R. S. *Master of Trinity College, Cambridge.*
Robert Smith, LL. B.
James Scawen, *of Maidwell,* Esq;
Edward Southwell, Esq;
Major St. Leger, *Fellow of St. Peter's College, Cambridge.*
The Hon. —— Sandys, Esq;
Mr. Thomas Simpson, F. R. S. *and Member of the Royal Academy of Sciences at Stockholm.*
Mr. Benjamin Smith.
Mr. Stephens, *of Deal.*
Mr. Smith, *Steward, and Land-Surveyor.*
Mr. Henry Smith, *Attorney at Law in Stamford.*
The Rev. Mr. John Sanderson.
The Rev. Mr. Anthony Sanderson.
Mr. William Smith.
Mr. Robert Sharman.
Mr. Thomas Saunder.

Thomas Truman, Esq;
Richard Tryce, Esq;
Trinity College, Cambridge, *for the Library.*
Mr. Tyson, C. C. C. C.
The Rev. Mr. Brownlowe Toller, 2 copies.
The Rev. Mr. Richard Turner, *Teacher of Geometry, Astronomy, and Philosophy, at Worcester.*
Mr. John Thistleton.
Mr. Daniel Tenand, *Writing-Master and Accomptant, in Islington.*

Mr. Thomas Taylor, *Writing-Master and Accomptant, in Shadwell.*
Mr. Tape.
Mr. John Thistlewood, *Land-Surveyor.*

Sir Philip Vavazor, *Knt.*

The Right Rev. Dr. John Thomas, *Lord Bishop of Winchester.*
Lady Charlotte Wentworth.
Matthew Wyldbore, Esq;
Thomas Whichcot, Esq; *Member of Parliament for the County of Lincoln.*
The Rev. Mr. Weston, *Rector of Thirfield, &c.*
The Rev. Mr. Waring, *Lucasian Professor of Mathematics in the University of Cambridge.*
Mr. Charlton Wyldbore.
Mr. Thomas White.
Mr. Benjamin Webb, *Writing-Master and Accomptant, in Bunhill-Row.*
Mr. Richard Willson.
Captain Henry Watson.
Captain Henry Willson.
Mr. Edmond Webster.
Mr. John Wing.
Mr. Tycho Wing, *Mathematical-Instrument-Maker, in London.*
Mr. J. Watts, *of Comb-Abbey,* 2 copies.
Mr. Henry Woolley, *Master of the Academy in Northampton.*
Mr. Richard Weston, *Teacher of the Mathematics, in Tanley.*
Mr. Jekyl Wilson.
Mr. Graham Wilkinson.
Mr. Joseph Wilkinson.

☞ Purchasers of this First Book will be esteemed Subscribers, and their Names (if they please to give them to the Author, or Publishers) shall be inserted in a complete List, to be printed with the Second Book:—In which Book, the Usefulness of this Analysis will be exemplified, in a variety of very curious Mechanical, and Physico-Geometrical, Enquiries.

J. LANDEN.

Milton, near Peterborough,
March 25, 1764.

THE

RESIDUAL ANALYSIS.

John White his Book Bennington near Boston Lincolnshire 7th May 1807.

CHAP. I.

Terms *and* Characters *explained.*

IT will be proper, in the firſt place, to give an explanation of certain *Terms* and *Characters*, which, in the courſe of this work, we ſhall have frequent occaſion to make uſe of:—accordingly, ſuch explanation is here premiſed.

I.

In any proceſs, a quantity that is conſidered as always retaining the ſame value, is called a *determinate* or *invariable* quantity. And a quantity that is not conſider'd as always retaining the ſame value, but may be taken of any value whatever, or of any value between certain limits, (ſome other quantity, or quantities, concerned therewith, at the ſame time, remaining invariable,) is called an *indeterminate* or *variable* quantity.

THE RESIDUAL

2.

An algebraic expreſſion compoſed, in any manner, of any power or powers of any variable quantity, with any invariable coefficients, is called a *function* of that quantity.

For inſtance, $a+bx^m+cx^n$ and $\frac{ex^p+\sqrt{f^2+x^2}}{gx^r}$ are functions of the variable quantity x: And, y being equal to any function of x, $\frac{axy+y\sqrt{x^2+y^2}}{x+b\sqrt{c+y}}$ may be alſo conſidered as a function of x.

If y be equal to any function of x, it is obvious that x will be equal to ſome function of y. And if y and z be equal to any functions of x, x and y may each be conſidered as a function of z; and x and z, each as a function of y.

3.

If, in any given expreſſion or function of x, wherein $\underset{,}{x}$ is not concerned, $\underset{,}{x}$ be ſubſtituted inſtead of x, the given expreſſion and that which reſults from ſuch ſubſtitution are called *ſimilar functions* of x and $\underset{,}{x}$ reſpectively.

For inſtance, $\frac{ex^p+\sqrt{f^2+x^2}}{gx^r}$ and $\frac{e\underset{,}{x}^p+\sqrt{f^2+\underset{,}{x}^2}}{g\underset{,}{x}^r}$ are ſimilar functions of x and $\underset{,}{x}$ reſpectively; the values of e, f, g, p, and r being independent of the values of x and $\underset{,}{x}$: And, y and $\underset{,}{y}$ being ſimilar functions of x and $\underset{,}{x}$ reſpectively, $ax^2y+by^m\sqrt{c^2+xy^n}$ may be conſidered as a function of x, and $a\underset{,}{x}^2\underset{,}{y}+b\underset{,}{y}^m\sqrt{c^2+\underset{,}{x}\underset{,}{y}^n}$ as a ſimilar function of $\underset{,}{x}$; or the ſame expreſſions ($ax^2y+by^m\sqrt{c^2+xy^n}$ and $a\underset{,}{x}^2\underset{,}{y}+b\underset{,}{y}^m\sqrt{c^2+\underset{,}{x}\underset{,}{y}^n}$) may be conſidered as ſimilar functions of y and $\underset{,}{y}$ reſpectively;

a,

a, b, c, m, and n being determinate, whilst x and $x_{,}$ are indeterminate.

4.

y being any function of x, and $y_{,}$ a similar function of $x_{,}$; the value of the quotient of $y - y_{,}$ divided by $x - x_{,}$, in the particular case when $x_{,}$ is equal to x, is called the *special value* of that quotient; and x and y are respectively named the *prime member of the divisor*, and the *prime member of the dividend*.

The said quotient, which algebraists commonly denote by $\frac{y - y_{,}}{x - x_{,}}$ or $\overline{y - y_{,}} \div \overline{x - x_{,}}$, we shall, for brevity sake, sometimes denote by $[x \mid y]$; and the special value thereof we shall express by $[x \perp y]$.

Moreover, $[x \perp y]$ and $[x_{,} \perp y_{,}]$ being similar functions of x and $x_{,}$, the special value of $\overline{[x \perp y] - [x_{,} \perp y_{,}]} \div \overline{x - x_{,}}$ will be denoted by $[x \perp\!\!\perp y]$:

And, in like manner, $[x \perp\!\!\perp y]$ and $[x_{,} \perp\!\!\perp y_{,}]$ being similar functions of x and $x_{,}$, the special value of $\overline{[x \perp\!\!\perp y] - [x_{,} \perp\!\!\perp y_{,}]} \div \overline{x - x_{,}}$ will be denoted by $[x \perp\!\!\perp\!\!\perp y]$: &c.

And the like is to be understood, when, for any quantities having the like relation, other letters or characters are substituted instead of x, $x_{,}$, y, and $y_{,}$.

Accordingly $[x_{,} \perp y_{,}]$ will denote the special value of $\overline{y_{,} - y_{,,}} \div \overline{x_{,} - x_{,,}}$, $y_{,}$ and $y_{,,}$ being similar functions of $x_{,}$ and $x_{,,}$ respectively: And, $[x_{,} \perp y_{,}]$ and $[x_{,,} \perp y_{,,}]$ being also similar functions of $x_{,}$

and $x_{\prime\prime}$, the special value of $\overline{[x_{\prime} \perp y_{\prime}] - [x_{\prime\prime} \perp y_{\prime\prime}]} \div \overline{x_{\prime} - x_{\prime\prime}}$ will be expressed by $[x_{\prime} \perp\!\!\perp y_{\prime}]$: &c.

So $[y \perp z]$ will signify the special value of $\overline{z - z_{\prime}} \div \overline{y - y_{\prime}}$, z and $z_{\prime}$ being similar functions of y and $y_{\prime}$ respectively:

And $[y \perp z]$ and $[y_{\prime} \perp z_{\prime}]$ being also similar functions of y and $y_{\prime}$, $[y \perp\!\!\perp z]$ will express the special value of $\overline{[y \perp z] - [y_{\prime} \perp z_{\prime}]} \div \overline{y - y_{\prime}}$: &c.

5.

Let it be noted, that, whenever y denotes any function of x; y, $y_{\prime}$, $y_{\prime\prime}$, $y_{\prime\prime\prime}$, &c. shall denote similar functions of x, $x_{\prime}$, $x_{\prime\prime}$, $x_{\prime\prime\prime}$, &c. respectively.

Likewise, whenever z denotes any function of y; z, $z_{\prime}$, $z_{\prime\prime}$, $z_{\prime\prime\prime}$, &c. shall denote similar functions of y, $y_{\prime}$, $y_{\prime\prime}$, $y_{\prime\prime\prime}$, &c. respectively. And the like is to be observed, with respect to other quantities.

6.

When any quantities v, x, y, z, &c. are concerned in any process, and the quantities $\underset{\cdot}{v}$, $\underset{\cdot}{x}$, $\underset{\cdot}{y}$, &c. or $[v \perp x]$, $[v \perp y]$, &c. are also concerned in the same process, it must be understood, that x, y, &c. are each consider'd as some function of v, though their being so related be not there particularly mentioned.

THE

THE

RESIDUAL ANALYSIS.

CHAP. II.

Of the Invention of Rules *necessary to faciliate Computations in this Analysis.*

IN making computations by the method taught in this Treatise, we shall frequently have occasion to assign the quotient of $F - \underset{\cdot}{F}$ divided by $x - \underset{\cdot}{x}$, F being some function of x; or, at least, we shall often find it necessary to have recourse to some Rule derived from a consideration of the relation between the special value of such quotient and the prime member of the correspondent dividend. Therefore, as the Books of Algebra hitherto published are deficient in respect to such division, it will here be proper to shew how such quotient may always be obtained; and to deduce therefrom such Rules as may be useful to facilitate computations in the succeeding Chapters.

I.

The Theorems which chiefly enable us to perform the division above mentioned are the following, viz. *

$$\frac{v^{\frac{m}{n}} - w^{\frac{m}{n}}}{v - w}$$

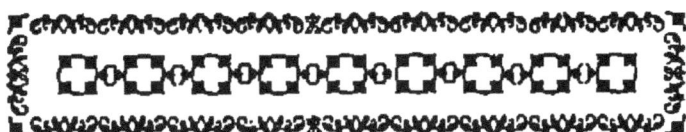

$$\frac{v^{\frac{m}{r}} - w^{\frac{m}{r}}}{v - w} = \frac{v^{m-1} + v^{m-2}w + v^{m-3}w^2 + v^{m-4}w^3 \ (m)}{v^{m-\frac{m}{r}} + v^{m-\frac{2m}{r}}w^{\frac{m}{r}} + v^{m-\frac{3m}{r}}w^{\frac{2m}{r}} + v^{m-\frac{4m}{r}}w^{\frac{3m}{r}} \ (r)} \text{*}$$

$$= v^{\frac{m}{r}-1} \times \frac{1 + \frac{w}{v} + \overline{\frac{w}{v}}\big|^2 + \overline{\frac{w}{v}}\big|^3 \ (m)}{1 + \overline{\frac{w}{v}}\big|^{\frac{m}{r}} + \overline{\frac{w}{v}}\big|^{\frac{2m}{r}} + \overline{\frac{w}{v}}\big|^{\frac{3m}{r}} \ (r)},$$

m and *r* being positive integers;

and $uw - u_{,}w_{,} = w \times \overline{u - u_{,}} + u_{,} \times \overline{w - w_{,}}$;

which, by an easy multiplication, will be found to be true.

From the first equation it follows, that $\frac{v^{-\frac{m}{r}} - w^{-\frac{m}{r}}}{v - w} = -$

$$\frac{v^{\frac{m}{r}} - w^{\frac{m}{r}}}{\overline{vw}\big|^{\frac{m}{r}} \times \overline{v - w}} \text{ is } = -v^{-1}w^{-\frac{m}{r}} \times \frac{1 + \frac{w}{v} + \overline{\frac{w}{v}}\big|^2 + \overline{\frac{w}{v}}\big|^3 \ (m)}{1 + \overline{\frac{w}{v}}\big|^{\frac{m}{r}} + \overline{\frac{w}{v}}\big|^{\frac{2m}{r}} + \overline{\frac{w}{v}}\big|^{\frac{3m}{r}} \ (r)}.$$

---

* This theorem may be investigated as follows.
It is well known, that

$$\frac{v^m - w^m}{v - w} \text{ is } = v^{m-1} + v^{m-2}w + v^{m-3}w^2 \ (m),$$

$$\text{and } \frac{a^r - b^r}{a - b} = a^{r-1} + a^{r-2}b + a^{r-3}b^2 \ (r),$$

m and *r* being positive Integers.

In the second equation write $v^{\frac{m}{r}}$ and $w^{\frac{m}{r}}$ instead of *a* and *b* respectively, and you will have

$$\frac{v^m - w^m}{v^{\frac{m}{r}} - w^{\frac{m}{r}}} = v^{m-\frac{m}{r}} + v^{m-\frac{2m}{r}}w^{\frac{m}{r}} + v^{m-\frac{3m}{r}}w^{\frac{2m}{r}} \ (r).$$

Then, from the first and third equations, it will appear by division, that

$$\frac{v^{\frac{m}{r}} - w^{\frac{m}{r}}}{v - w} \text{ is } = \frac{v^{m-1} + v^{m-2}w \ (m)}{v^{m-\frac{m}{r}} + v^{m-\frac{2m}{r}}w^{\frac{m}{r}} \ (r)}:$$

Which being so obvious, it is matter of surprize to me, that Algebraists have not before observed it, and shewn its singular use in Analytics.

EXAMPLE

EXAMPLE I. Taking $\frac{m}{r} = \frac{4}{3}$ we have

$$\overline{v^{\frac{4}{3}} - w^{\frac{4}{3}}} \div \overline{v - w} = v^{\frac{1}{3}} \times \frac{1 + \frac{w}{v} + \overline{\frac{w}{v}}|^{2} + \overline{\frac{w}{v}}|^{3}}{1 + \overline{\frac{w}{v}}|^{\frac{4}{3}} + \overline{\frac{w}{v}}|^{\frac{8}{3}}}.$$

Whence it is evident, that, when w is $= v$, the quotient of $v^{\frac{4}{3}} - w^{\frac{1}{4}}$ divided by $v - w$ is equal to $\frac{4}{3}v^{\frac{1}{3}}$.

Moreover, $\frac{4}{3}$ being $= 1.3333$ &c.

$$\overline{v^{\frac{4}{3}} - w^{\frac{4}{3}}} \div \overline{v - w} \text{ is} = v^{\frac{1}{3}} \times \frac{1 + \frac{w}{v} + \overline{\frac{w}{v}}|^{2} + \overline{\frac{w}{v}}|^{3} \ (13333)}{1 + \overline{\frac{w}{v}}|^{\frac{4}{3}} + \overline{\frac{w}{v}}|^{\frac{8}{3}} + \overline{\frac{w}{v}}|^{\frac{12}{3}} \ (10000)} \text{ nearly;}$$

$$\text{or, more nearly, } = v^{\frac{1}{3}} \times \frac{1 + \frac{w}{v} + \overline{\frac{w}{v}}|^{2} + \overline{\frac{w}{v}}|^{3} \ (133333)}{1 + \overline{\frac{w}{v}}|^{\frac{4}{3}} + \overline{\frac{w}{v}}|^{\frac{4}{3}} + \overline{\frac{w}{v}}|^{\frac{12}{3}} \ (100000)};$$

&c.

Hence it is likewise evident, that, when w is $= v$, the quotient of $v^{\frac{4}{3}} - w^{\frac{4}{3}}$ divided by $v - w$ is equal to $\frac{4}{3}v^{\frac{1}{3}}$: For, the ratio of 1333 &c. to 1000 &c. being as 1.333 &c. to 1, the ultimate value of $\frac{1+1+1+1 \ (1333 \text{ \&c.})}{1+1+1+1 \ (1000 \text{ \&c.})}$ (the value of $\frac{1 + \frac{w}{v} + \overline{\frac{w}{v}}|^{2} \ (1333 \text{ \&c.})}{1 + \overline{\frac{w}{v}}|^{\frac{4}{3}} + \overline{\frac{w}{v}}|^{\frac{8}{3}} \ (1000 \text{ \&c.})}$ when w is equal to v) is manifestly equal to $\frac{4}{3}$, the quantity from which (by division) 1.333 &c. is obtained.

EXAMPLE

EXAMPLE II. Taking $\frac{m}{r} = \sqrt{2} = 1.4142$ &c. we have

$$\overline{v^{\sqrt{2}} - w^{\sqrt{2}}} \div \overline{v - w} = v^{\sqrt{2}-1} \times \frac{1 + \frac{w}{v} + \overline{\tfrac{w}{v}}|^{2} + \overline{\tfrac{w}{v}}|^{3} \quad (1414)}{1 + \overline{\tfrac{w}{v}}|^{\sqrt{2}} + \overline{\tfrac{w}{v}}|^{2\sqrt{2}} + \overline{\tfrac{w}{v}}|^{3\sqrt{2}} \quad (1000)} \text{ nearly;}$$

$$\text{or, more nearly, } = v^{\sqrt{2}-1} \times \frac{1 + \frac{w}{v} + \overline{\tfrac{w}{v}}|^{2} + \overline{\tfrac{w}{v}}|^{3} \quad (14142)}{1 + \overline{\tfrac{w}{v}}|^{\sqrt{2}} + \overline{\tfrac{w}{v}}|^{2\sqrt{2}} + \overline{\tfrac{w}{v}}|^{3\sqrt{2}} \quad (10000)};$$

&c.

Whence it is evident, that, when w is $= v$, the quotient of $v^{\sqrt{2}} - w^{\sqrt{2}}$ divided by $v - w$ is equal to $\sqrt{2} \times v^{\sqrt{2}-1}$: For the ratio of 14142 &c. to 10000 &c. being as 1.4142 &c. to 1, it is manifest, that the ultimate value of $\frac{1+1+1+1 \ (14142 \text{ \&c.})}{1+1+1+1 \ (10000 \text{ \&c.})}$ (the value of $\frac{1 + \frac{w}{v} + \overline{\tfrac{w}{v}}|^{2} + \overline{\tfrac{w}{v}}|^{3} \quad (14142 \text{ \&c.})}{1 + \overline{\tfrac{w}{v}}|^{\sqrt{2}} + \overline{\tfrac{w}{v}}|^{2\sqrt{2}} + \overline{\tfrac{w}{v}}|^{3\sqrt{2}} \quad (10000 \text{ \&c.})}$ when w is equal to v) is equal to $\sqrt{2}$, the quantity from which (by extracting the root) 1.4142 &c. is derived.

COROLLARY I. Seeing $\frac{1 + \frac{w}{v} + \overline{\tfrac{w}{v}}|^{2} + \overline{\tfrac{w}{v}}|^{3} \quad (m)}{1 + \overline{\tfrac{w}{v}}|^{\frac{m}{r}} + \overline{\tfrac{w}{v}}|^{\frac{2m}{r}} + \overline{\tfrac{w}{v}}|^{\frac{3m}{r}} \quad (r)}$ is equal to $\frac{m}{r}$, when w is $= v$; it is evident, that the special value of $\overline{v^{\frac{m}{r}} - w^{\frac{m}{r}}} \div \overline{v - w}$ is equal to $\frac{m}{r} v^{\frac{m}{r} - 1}$, whether $\frac{m}{r}$ be positive or negative.

Hence, by substituting p, x, and $\underset{,}{x}$, instead of $\frac{m}{r}$, v, and w respectively; it appears, that, p being either positive or negative, $[x \perp x^{p}]$ (the special value of $\overline{x^{p} - \underset{,}{x}^{p}} \div \overline{x - \underset{,}{x}}$) is $= px^{p-1}$.

COROLLARY

COROLLARY II. $\overline{k + av^{\frac{m}{r}} + bv^{\frac{n}{s}} \&c.} - \overline{k + w^{\frac{m}{r}} + bw^{\frac{n}{s}} \&c.}$

being $= a \times \overline{v^{\frac{m}{r}} - w^{\frac{m}{r}}} + b \times \overline{v^{\frac{n}{s}} - w^{\frac{n}{s}}} \&c.$

$$\overline{\overline{av^{\frac{m}{r}} + bv^{\frac{n}{s}} \&c.} - \overline{aw^{\frac{m}{r}} + bw^{\frac{n}{s}} \&c.}} \div \overline{v - w}$$

$$\text{is} = av^{\frac{m}{r} - 1} \times M + bv^{\frac{n}{s} - 1} \times N \ \&c.$$

and $\overline{\overline{av^{-\frac{m}{r}} + bv^{-\frac{n}{s}} \&c.} - \overline{aw^{-\frac{m}{r}} + bw^{-\frac{n}{s}} \&c.}} \div \overline{v - w}$

$$= -av^{-1}w^{-\frac{m}{r}} \times M - bv^{-1}w^{-\frac{n}{s}} \times N \ \&c.$$

$$M \text{ being} = \frac{1 + \frac{w}{v} + \overline{\frac{w}{v}}\Big|^{2} + \overline{\frac{w}{v}}\Big|^{3} \quad (m)}{1 + \overline{\frac{w}{v}}\Big|^{\frac{m}{r}} + \overline{\frac{w}{v}}\Big|^{\frac{2m}{r}} + \overline{\frac{w}{v}}\Big|^{\frac{3m}{r}} \quad (r)},$$

$$\text{and } N = \frac{1 + \frac{w}{v} + \overline{\frac{w}{v}}\Big|^{2} + \overline{\frac{w}{v}}\Big|^{3} \quad (n)}{1 + \overline{\frac{w}{v}}\Big|^{\frac{n}{s}} + \overline{\frac{w}{v}}\Big|^{\frac{2n}{s}} + \overline{\frac{w}{v}}\Big|^{\frac{3n}{s}} \quad (s)},$$

&c. &c.

Moreover,

$[x \perp \overline{k + ax^{p} + bx^{q} \&c.}]$ is, in general, $= pax^{p-1} + qbx^{q-1} \&c.$

2.

Suppose $\overline{z^{\frac{m}{r}} - z_{,}^{\frac{m}{r}}} \div \overline{z - z_{,}} = F$, and $\overline{z - z_{,}} \div \overline{x - x_{,}} = f$:

Then, $z - z_{,}$ being $= fx - fx_{,}$, $\overline{z^{\frac{m}{r}} - z_{,}^{\frac{m}{r}}} \div \overline{fx - fx_{,}}$ will

be $= F$, and $\overline{z^{\frac{m}{r}} - z_{,}^{\frac{m}{r}}} \div \overline{x - x_{,}} = Ff$.

EXAMPLE. Let z be supposed $= ax + x^{\frac{2}{3}}$, and $z_{,} = ax_{,} + x_{,}^{\frac{2}{3}}$:

then,

then, F being (by what is said above) $= z^{\frac{m}{r}-1} \times \dfrac{1 + \frac{z_{,}}{z} + \overline{\frac{z_{,}}{z}}\big|^{2} \cdots (m)}{1 + \overline{\frac{z_{,}}{z}}\big|^{\frac{m}{r}} + \overline{\frac{z_{,}}{z}}\big|^{\frac{2m}{r}} \cdots (r)}$,

and $f = a + x^{-\frac{1}{3}} \times \dfrac{1 + \frac{x_{,}}{x}}{1 + \overline{\frac{x_{,}}{x}}\big|^{\frac{1}{3}} + \overline{\frac{x_{,}}{x}}\big|^{\frac{2}{3}}}$, $\overline{z^{\frac{m}{r}} - z_{,}^{\frac{m}{r}}} \div \overline{x - x_{,}}$

will be

$$= z^{\frac{m}{r}-1} \times \frac{1 + \frac{z_{,}}{z} + \overline{\frac{z_{,}}{z}}\big|^{2} \cdots (m)}{1 + \overline{\frac{z_{,}}{z}}\big|^{\frac{m}{r}} + \overline{\frac{z_{,}}{z}}\big|^{\frac{2m}{r}} \cdots (r)} \times \overline{a + x^{-\frac{1}{3}} \times \frac{1 + \frac{x_{,}}{x}}{1 + \overline{\frac{x_{,}}{x}}\big|^{\frac{1}{3}} + \overline{\frac{x_{,}}{x}}\big|^{\frac{2}{3}}}}.$$

Corollary. When $x_{,}$ is $= x$, $z_{,}$ will be $= z$, and $f = [x \perp z]$, z being any function of x; moreover, by Cor. 1. of the preceding Article, F will be $= pz^{p-1}$, p being written instead of $\frac{m}{r}$.

Therefore $[x \perp z^{p}]$ will be $= pz^{p-1} \times [x \perp z]$.

Example I. Suppose $z = a + bx^{m} + cx^{n}$ &c. Then, $[x \perp z]$, by the preceding Article, being $= bmx^{m-1} + cnx^{n-1}$ &c. $[x \perp \overline{a + bx^{m} + cx^{n} \text{ \&c.}}|^{p}]$ will be $=$
$p \times \overline{a + bx^{m} + cx^{n} \text{ \&c.}}|^{p-1} \times \overline{bmx^{m-1} + cnx^{n-1} \text{ \&c.}}$

Supposing the coefficients c, d, &c. each $= 0$, we have

$$[x \perp \overline{a + bx^{m}}|^{p}] = bmp \times \overline{a + bx^{m}}|^{p-1} \times x^{m-1}.$$

Example

EXAMPLE II. Let z be supposed $= cx^n + \overline{a + bx^m}|^p$. Then $\overline{cx^n + \overline{a+bx^m}|^p - c\underset{,}{x}^n + \overline{a + b\underset{,}{x}^m}|^p}$ being $= c.\overline{x^n - \underset{,}{x}^n} + \overline{\overline{a + bx^m}|^p - \overline{a + b\underset{,}{x}^m}|^p}$, $[x \perp cx^n + \overline{a + bx^m}|^p]$ will be $=$ $[x \perp cx^n] + [x \perp \overline{a + bx^m}|^p] = cnx^{n-1} + bmp \times \overline{a + bx^m}|^{p-1} \times x^{m-1}$, and $[x \perp \overline{cx^n + \overline{a + bx^m}|^p}|^q] = q \times \overline{cx^n + \overline{a + bx^m}|^p}|^{q-1} \times \overline{cnx^{n-1} + bmp \times \overline{a + bx^m}|^{p-1} \times x^{m-1}}$.

3.

Seeing $uw - \underset{,}{u}\underset{,}{w}$ is $= w \times \overline{u - \underset{,}{u}} + \underset{,}{u} \times \overline{w - \underset{,}{w}}$, as we have before observed; it must follow, that $\overline{uw - \underset{,}{u}\underset{,}{w}} \div \overline{v - \underset{,}{v}}$ is $= w \times \frac{u - \underset{,}{u}}{v - \underset{,}{v}} + \underset{,}{u} \times \frac{w - \underset{,}{w}}{v - \underset{,}{v}}$.

EXAMPLE. By taking u, w, $\underset{,}{u}$, and $\underset{,}{w}$ respectively equal to $\overline{a^2 - v^2}|^{\frac{1}{2}}$, $a^{\frac{1}{2}} + v^{\frac{1}{2}}$, $\overline{a^2 - \underset{,}{v}^2}|^{\frac{1}{2}}$, and $a^{\frac{1}{2}} + \underset{,}{v}^{\frac{1}{2}}$, we have $\overline{\overline{a^2 - v^2}|^{\frac{1}{2}} \times \overline{a^{\frac{1}{2}} + v^{\frac{1}{2}}} - \overline{a^2 - \underset{,}{v}^2}|^{\frac{1}{2}} \times \overline{a^{\frac{1}{2}} + \underset{,}{v}^{\frac{1}{2}}}} \div \overline{v - \underset{,}{v}} =$ $- \overline{a^{\frac{1}{2}} + v^{\frac{1}{2}}} \times \frac{v + \underset{,}{v}}{\sqrt{a^2 - v^2} + \sqrt{a^2 - \underset{,}{v}^2}} + \sqrt{a^2 - \underset{,}{v}^2} \times \frac{1}{v^{\frac{1}{2}} + \underset{,}{v}^{\frac{1}{2}}}$; $\overline{\overline{a^2 - v^2}|^{\frac{1}{2}} - \overline{a^2 - \underset{,}{v}^2}|^{\frac{1}{2}}} \div \overline{v - \underset{,}{v}}$ (by what is said above) being equal to $- \frac{v + \underset{,}{v}}{\sqrt{a^2 - v^2} + \sqrt{a^2 - \underset{,}{v}^2}}$, and $\overline{a^{\frac{1}{2}} + v^{\frac{1}{2}} - \overline{a^{\frac{1}{2}} + \underset{,}{v}^{\frac{1}{2}}}} \div \overline{v - \underset{,}{v}}$ $(= \overline{v^{\frac{1}{2}} - \underset{,}{v}^{\frac{1}{2}}} \div \overline{v - \underset{,}{v}})$ equal to $\frac{1}{v^{\frac{1}{2}} + \underset{,}{v}^{\frac{1}{2}}}$.

COROLLARY. When $\underset{\cdot}{v}$ is $= v$, $\underset{\cdot}{u}$ and $\underset{\cdot}{w}$ will be equal to u and w respectively, these last quantities being any functions of v. It is manifest therefore, that

$$[v \perp uw] \text{ will be } = w\,[v \perp u] + u\,[v \perp w].$$

EXAMPLE I. Supposing u and w to be respectively equal to x^m and y^n, (*i. e.* supposing x and y to be equal to any functions of v,) we have $[v \perp x^m y^n] = y^n \times [v \perp x^m] + x^m \times [v \perp y^n]$ $= mx^{m-1}y^n\,[v \perp x] + nx^m y^{n-1}\,[v \perp y]$.

EXAMPLE II. Supposing v, u, and w equal to x, cx^m, and $\overline{a+bx^r}|^{n}$ respectively, we have

$$[x \perp cx^m \times \overline{a+bx^r}|^{n}] = \overline{a+bx^r}|^{n} \times [x \perp cx^m] + cx^m \times [x \perp \overline{a+bx^r}|^{n}]$$
$$= cm \times \overline{a+bx^r}|^{n} \times x^{m-1} + cnx^m \times \overline{a+bx^r}|^{n-1} \times [x \perp a + bx^r]$$
$$= cm \times \overline{a+bx^r}|^{n} \times x^{m-1} + bcnr \times \overline{a+bx^r}|^{n-1} \times x^{m+r-1}.$$

EXAMPLE III. Taking u and w respectively equal to $\overline{a+bx^n}|^{m}$ and $\overline{c+dx^r}|^{-1}$, we find

$$[v \perp \overline{a+bx^n}|^{m} \times \overline{c+dx^r}|^{-1}] = \overline{c+dx^r}|^{-1} \times [v \perp \overline{a+bx^n}|^{m}]$$
$$+ \overline{a+bx^n}|^{m} \times [v \perp \overline{c+dx^r}|^{-1}] = m \times \overline{c+dx^r}|^{-1} \times \overline{a+bx^n}|^{m-1}$$
$$\times [v \perp a+bx^n] - \overline{a+bx^n}|^{m} \times \overline{c+dx^r}|^{-2} \times [v \perp c+dx^r] =$$
$$bmn \times \overline{c+dx^r}|^{-1} \times \overline{a+bx^n}|^{m-1} \times x^{n-1}[v \perp x] - dr \times \overline{a+bx^n}|^{m} \times$$
$\overline{c+dx^r}|^{-2} \times x^{r-1}[v \perp x]$; a, b, c, d, m, n, and r being determinate, whilst x is indeterminate.

EXAMPLE IV. y being any function of x, we have

$$[x \perp \overline{a+bx^n}|^{m} \times y^r] = y^r \times [x \perp \overline{a+bx^n}|^{m}] + \overline{a+bx^n}|^{m} \times [x \perp y^r]$$
$$= my^r \times \overline{a+bx^n}|^{m-1} \times [x \perp a+bx^n] + ry^{r-1} \times \overline{a+bx^n}|^{m} \times [x \perp y]$$
$$= bmnx^{n-1}y^r \times \overline{a+bx^n}|^{m-1} + ry^{r-1} \times \overline{a+bx^n}|^{m} \times [x \perp y].$$

EXAMPLE

EXAMPLE V.

$[x \perp z[x \perp z]]$ is $= [x \perp z]^2 + z[x \perp z]$,
$[x \perp z[x \perp z]] = [x \perp z] \times [x \perp z] + z[x \perp z]$,
$[x \perp y[x \perp z]] = [x \perp y] \times [x \perp z] + y[x \perp z]$, and
$[x \perp [x \perp y] \times [x \perp z]] = [x \perp y] \times [x \perp z] + [x \perp y] \times [x \perp z]$.
y and z being any functions of x.

4.

$uw - \underset{,\,,}{uw}$ being (as is observed in the last Article) $= w \times \overline{u - \underset{,}{u}} + \underset{,}{u} \times \overline{w - \underset{,}{w}} = A$; $uwx - \underset{,\,,\,,}{uwx}$ will be $= xA + \underset{,\,,}{uw} \times \overline{x - \underset{,}{x}} = wx \times \overline{u - \underset{,}{u}} + x\underset{,}{u} \times \overline{w - \underset{,}{w}} + \underset{,\,,}{uw} \times \overline{x - \underset{,}{x}} = B$; and $uwxy - \underset{,\,,\,,\,,}{uwxy} = yB + \underset{,\,,\,,}{uwx} \times \overline{y - \underset{,}{y}} = wxy \times \overline{u - \underset{,}{u}} + xy\underset{,}{u} \times \overline{w - \underset{,}{w}} + y\underset{,\,,}{uw} \times \overline{x - \underset{,}{x}} + \underset{,\,,\,,}{uwx} \times \overline{y - \underset{,}{y}} = C$; &c.

It is evident therefore, that $uwx\,(n) - \underset{,\,,\,,}{uwx}\,(n)$ is $= wx\,(n-1) \times \overline{u - \underset{,}{u}} + xy\,(n-2) \times \underset{,}{u} \times \overline{w - \underset{,}{w}} + yz\,(n-3) \times \underset{,\,,}{uw} \times \overline{x - \underset{,}{x}} + (n)$; and, consequently,

$$\overline{uwx\,(n) - \underset{,\,,\,,}{uwx}\,(n)} \div \overline{v - \underset{,}{v}} = wx\,(n-1) \times \frac{u - \underset{,}{u}}{v - \underset{,}{v}} + xy\,(n-2) \times \underset{,}{u} \times \frac{w - \underset{,}{w}}{v - \underset{,}{v}} + yz\,(n-3) \times \underset{,\,,}{uw} \times \frac{x - \underset{,}{x}}{v - \underset{,}{v}} + (n).$$

EXAMPLE. Taking u, w, x, $\underset{,}{u}$, $\underset{,}{w}$, and $\underset{,}{x}$ equal to $v^{\frac{m}{n}}$, $\overline{a+v}|^{\frac{r}{s}}$, $\overline{b+v}|^{\frac{p}{q}}$, $\underset{,}{v}^{\frac{m}{n}}$, $\overline{a+\underset{,}{v}}|^{\frac{r}{s}}$, and $\overline{b+\underset{,}{v}}|^{\frac{p}{q}}$, respectively; we have $\overline{v^{\frac{m}{n}} \times \overline{a+v}|^{\frac{r}{s}} \times \overline{b+v}|^{\frac{p}{q}} - \underset{,}{v}^{\frac{m}{n}} \times \overline{a+\underset{,}{v}}|^{\frac{r}{s}} \times \overline{b+\underset{,}{v}}|^{\frac{p}{q}}}$

$\div$

$$\div\ \overline{v - v_{,}} = \overline{a+v}\big|^{\frac{1}{3}} \times \overline{b+v}\big|^{\frac{1}{3}} \times v^{-\frac{2}{3}} \times \frac{1}{1 + \overline{\frac{v_{,}}{v}}\big|^{\frac{1}{3}} + \overline{\frac{v_{,}}{v}}\big|^{\frac{2}{3}}} +$$

$$\overline{b+v}\big|^{\frac{1}{3}} \times v_{,}^{\frac{1}{3}} \times \overline{a+v}\big|^{-\frac{2}{3}} \times \frac{1}{1 + \overline{\frac{a+v_{,}}{a+v}}\big|^{\frac{1}{3}}} + v_{,}^{\frac{1}{3}} \times \overline{a+v_{,}}\big|^{\frac{1}{3}} \times$$

$$\overline{b+v}\big|^{-\frac{2}{3}} \times \frac{1 + \frac{b+v_{,}}{b+v}}{1 + \overline{\frac{b+v_{,}}{b+v}}\big|^{\frac{1}{3}} + \overline{\frac{b+v_{,}}{b+v}}\big|^{\frac{2}{3}}};\quad \overline{v^{\frac{1}{3}} - v_{,}^{\frac{1}{3}}} \div \overline{v - v_{,}}$$ (by what is said above) being equal to $v^{-\frac{2}{3}} \times \frac{1}{1 + \overline{\frac{v_{,}}{v}}\big|^{\frac{1}{3}} + \overline{\frac{v_{,}}{v}}\big|^{\frac{2}{3}}}$,

$\overline{\overline{a+v}\big|^{\frac{1}{3}} - \overline{a+v_{,}}\big|^{\frac{1}{3}}} \div \overline{v - v_{,}}$ equal to $\overline{a+v}\big|^{-\frac{2}{3}} \times \frac{1}{1 + \overline{\frac{a+v_{,}}{a+v}}\big|^{\frac{1}{3}}}$,

and $\overline{\overline{b+v}\big|^{\frac{1}{3}} - \overline{b+v_{,}}\big|^{\frac{1}{3}}} \div \overline{v - v_{,}}$ equal to

$$\overline{b+v}\big|^{-\frac{2}{3}} \times \frac{1 + \frac{b+v_{,}}{b+v}}{1 + \overline{\frac{b+v_{,}}{b+v}}\big|^{\frac{1}{3}} + \overline{\frac{b+v_{,}}{b+v}}\big|^{\frac{2}{3}}}.$$

COROLLARY. When $v_{,}$ is equal to v; $u_{,}$, $w_{,}$, $x_{,}$, $y_{,}$, &c. will be respectively equal to u, w, x, y, &c. these last quantities being any functions of v. It is evident therefore, that $[v \perp uwx\,(n)]$ will be $= wxy\,(n-1) \times [v \perp u] + uxy\,(n-1) \times [u \perp w] + uwy\,(n-1) \times [v \perp x] + (n)$.

EXAMPLE. Writing u^p, w^q, x^r, y^s; instead of u, w, x, and y respectively: we have $[v \perp u^p w^q x^r y^s] = w^q x^r y^s \times [v \perp u^p] +$

u^p

$u^p x^r y^s \times [v \perp w^q] + u^p w^q y^s \times [v \perp x^r] + u^p w^q x^r \times [v \perp y^s] = p w^q x^r y^s u^{p-1} \times [v \perp u] + q u^p x^r y^s w^{q-1} \times [v \perp w] + r u^p w^q y^s x^{r-1} \times [v \perp x] + s u^p w^q x^r y^{s-1} \times [v \perp y]$; u, w, x, and y being any functions of v.

5.

Suppose $\overline{F - F_{,}} \div \overline{x - x_{,}}, = Q$, and $\overline{f - f_{,}} \div \overline{x - x_{,}} = q$: Then, $F - F_{,}$ being $= Q \times \overline{x - x_{,}}$, and $f - f_{,} = q \times \overline{x - x_{,}}$,

$$\overline{F - F_{,}} \div \overline{f - f_{,}} \text{ is manifestly } = \frac{Q}{q}.$$

EXAMPLE. If F, f, $F_{,}$, and $f_{,}$ be equal to $\overline{ax + x^2}|^{\frac{2}{3}}$, $x^{\frac{1}{2}}$, $\overline{ax_{,} + x_{,}^2}|^{\frac{2}{3}}$, and $x_{,}^{\frac{1}{2}}$ respectively; $\overline{F - F_{,}} \div \overline{f - f_{,}}$ will, by what is done above, be

$$= \overline{ax + x^2}|^{-\frac{1}{3}} \times \frac{1 + \frac{ax_{,} + x_{,}^2}{ax + x^2}}{1 + \overline{\frac{ax_{,} + x_{,}^2}{ax + x^2}}|^{\frac{1}{3}} + \overline{\frac{ax_{,} + x_{,}^2}{ax + x^2}}|^{\frac{2}{3}}} \times \overline{a + x_{,} + x} \div \frac{1}{x^{\frac{1}{2}} + x_{,}^{\frac{1}{2}}}.$$

COROLLARY. F and f being any functions of x, the value of the quotient or fraction $\frac{F - F_{,}}{f - f_{,}}$, when $x_{,}$ is $= x$, (*i. e.* when both numerator and denominator vanish together,) is $= \frac{[x \perp F]}{[x \perp f]}$.

EXAMPLE.

EXAMPLE. If F and f be equal to $\overline{a^2+x^2}|^{\frac{3}{2}}$ and $\overline{a+x}|^{\frac{3}{2}}$ respectively, the value of $\frac{F-F_{,}}{f-f_{,}}$ $\left(=\frac{\overline{a^2+x^2}|^{\frac{3}{2}}-\overline{a^2+x_{,}^2}|^{\frac{3}{2}}}{\overline{a+x}|^{\frac{3}{2}}-\overline{a+x_{,}}|^{\frac{3}{2}}}\right)$, when $x_{,}$ is $= x$, is equal to $\frac{3x\times\overline{a^2+x^2}|^{\frac{1}{2}}}{\frac{3}{2}\times\overline{a+x}|^{\frac{1}{2}}} = 2x\times\overline{\frac{a^2+x^2}{a+x}}|^{\frac{1}{2}}$.

6.

N and D being algebraic expressions so composed of x and other quantities, that each of those expressions becomes equal to *Nothing*, when x is equal to some certain quantity k; it is obvious, that $N_{,}$ and $D_{,}$ (the expressions which result by writing $x_{,}$ instead of x in N and D respectively) will both vanish when $x_{,}$ is equal to k. Therefore, if, in the quotient of $\overline{N-N_{,}}$ divided by $\overline{D-D_{,}}$, $x_{,}$ be taken equal to k, the resulting expression will be equal to $\frac{N}{D}$, let x be what it will. Now, by what is shewn above, the quotient of $\overline{N-N_{,}}$ divided by $\overline{D-D_{,}}$ $\left(=\frac{\overline{N-N_{,}}\div\overline{x-x_{,}}}{\overline{D-D_{,}}\div\overline{x-x_{,}}}\right)$ may always be readily assigned in terms which shall not vanish when x and $x_{,}$ are each equal to k, unless the value of that quotient be then $=0$: consequently, by assigning such quotient, the value of $\frac{N}{D}$ will be had, as well when N and D are each equal to *Nothing*, as in any other case.

EXAMPLE I. *To divide* $\sqrt{2a^2 + 2x^2} - 2a^{\frac{2}{3}}x^{\frac{1}{3}}$ *by* $x - a$.

The dividend and divisor both vanishing when x is equal to a, it is obvious, that, if, in the quotient of

$\overline{\sqrt{2a^2 + 2x^2} - 2a^{\frac{2}{3}}x^{\frac{1}{3}} - \sqrt{2a^2 + 2\underset{,}{x}^2} - 2a^{\frac{2}{3}}\underset{,}{x}^{\frac{1}{3}}}$ divided by $\overline{x - a} - \overline{\underset{,}{x} - a}$, $\underset{,}{x}$ be taken equal to a, the resulting expression will be equal to $\frac{\sqrt{2a^2 + 2x^2} - 2a^{\frac{2}{3}}x^{\frac{1}{3}}}{x - a}$, let x be what it will.

Now $\frac{\sqrt{2a^2 + 2x^2} - 2a^{\frac{2}{3}}x^{\frac{1}{3}} - \sqrt{2a^2 + 2\underset{,}{x}^2} - 2a^{\frac{2}{3}}\underset{,}{x}^{\frac{1}{3}}}{x - a - \underset{,}{x} - a}$ is manifestly

$= \frac{\sqrt{2a^2 + 2x^2} - \sqrt{2a^2 + 2\underset{,}{x}^2}}{x - \underset{,}{x}} - 2a^{\frac{2}{3}} \times \frac{x^{\frac{1}{3}} - \underset{,}{x}^{\frac{1}{3}}}{x - \underset{,}{x}}$; which, by what is done above, is $= \frac{2 \times \overline{x + \underset{,}{x}}}{\sqrt{2a^2 + 2x^2} + \sqrt{2a^2 + 2\underset{,}{x}^2}} - \frac{2a^{\frac{2}{3}}}{x^{\frac{2}{3}} + x^{\frac{1}{3}}\underset{,}{x}^{\frac{1}{3}} + \underset{,}{x}^{\frac{2}{3}}}$.

Consequently, by taking $\underset{,}{x}$ equal to a, we have

$$\frac{\sqrt{2a^2 + 2x^2} - 2a^{\frac{2}{3}}x^{\frac{1}{3}}}{x - a} = \frac{2 \times \overline{a + x}}{2a + \sqrt{2a^2 + 2x^2}} - \frac{2a^{\frac{2}{3}}}{a^{\frac{2}{3}} + a^{\frac{1}{3}}x^{\frac{1}{3}} + x^{\frac{2}{3}}}.$$

Whence it is evident, that, when x is equal to a, the quotient of $\sqrt{2a^2 + 2x^2} - 2a^{\frac{2}{3}}x^{\frac{1}{3}} \div \overline{x - a}$ is $= \frac{1}{3}$.

EXAMPLE II. *Suppose* N *and* D *equal to* $\sqrt{2k^3x - x^4} - \sqrt{kx^3}$ *and* $2x - \overline{7k^3 + x^3}|^{\frac{1}{3}}$ *respectively.*

Then $\overline{\sqrt{2k^3x - x^4} - \sqrt{kx^3} - \sqrt{2k^3\underset{,}{x} - \underset{,}{x}^4} - \sqrt{k\underset{,}{x}^3}} \div \overline{x - \underset{,}{x}}$ being $= \overline{\sqrt{2k^3x - x^4} - \sqrt{2k^3\underset{,}{x} - \underset{,}{x}^4}} \div \overline{x - \underset{,}{x}} -$

$$\overline{\sqrt{kx^3} - \sqrt{kx_{,}^3}} \div \overline{x - x_{,}} = \frac{2k^3 - x^3 - x^2x_{,} - xx_{,}^2 - x_{,}^3}{\sqrt{2k^3x - x^4} + \sqrt{2k^3x_{,} - x_{,}^4}} - k^{\frac{1}{2}}x^{\frac{1}{2}} \times$$

$$\frac{\left(1 + \frac{x_{,}}{x} + \frac{x_{,}}{x}\right)^{2}}{\left(1 + \frac{x_{,}}{x}\right)^{\frac{1}{2}}},$$ and $$\overline{2x - \left(7k^3 + x^3\right)^{\frac{1}{3}} - 2x_{,} - \left(7k^3 + x_{,}^3\right)^{\frac{1}{3}}} \div$$

$$\overline{x - x_{,}} = \overline{2x - 2x_{,}} \div \overline{x - x_{,}} - \overline{\left(7k^3 + x^3\right)^{\frac{1}{3}} - \left(7k^3 + x_{,}^3\right)^{\frac{1}{3}}}$$

$$\div \overline{x - x_{,}} = 2 - \left(7k^3 + x^3\right)^{-\frac{2}{3}} \times \frac{x^2 + xx_{,} + x_{,}^2}{1 + \left(\frac{7k^3 + x_{,}^3}{7k^3 + x^3}\right)^{\frac{1}{3}} + \left(\frac{7k^3 + x_{,}^3}{7k^3 + x^3}\right)^{\frac{2}{3}}},$$

$$\frac{\overline{N - N_{,}} \div \overline{x - x_{,}}}{\overline{D - D_{,}} \div \overline{x - x_{,}}}$$ will be $=$

$$\frac{\dfrac{2k^3 - x^3 - x^2x_{,} - xx_{,}^2 - x_{,}^3}{\sqrt{2k^3x - x^4} + \sqrt{2k^3x_{,} - x_{,}^4}} - k^{\frac{1}{2}}x^{\frac{1}{2}} \times \dfrac{\left(1 + \frac{x_{,}}{x} + \frac{x_{,}}{x}\right)^{2}}{\left(1 + \frac{x_{,}}{x}\right)^{\frac{1}{2}}}}{2 - \dfrac{x^2 + xx_{,} + x_{,}^2}{\left(7k^3 + x^3\right)^{\frac{2}{3}} + \left(7k^3 + x^3\right)^{\frac{1}{3}} \times \left(7k^3 + x_{,}^3\right)^{\frac{1}{3}} + \left(7k^3 + x_{,}^3\right)^{\frac{2}{3}}}}.$$

From whence, by taking $x_{,}$ equal to k, we get $\frac{N}{D} =$

$$\frac{\dfrac{k^3 - x^3 - kx^2 - k^2x}{\sqrt{2k^3x - x^4} + k^2} - \left(kx\right)^{\frac{1}{2}} \times \dfrac{\left(1 + \frac{k}{x} + \frac{k}{x}\right)^{2}}{\left(1 + \frac{k}{x}\right)^{\frac{1}{2}}}}{2 - \dfrac{x^2 + kx + k^2}{\left(7k^3 + x^3\right)^{\frac{2}{3}} + 2k \times \left(7k^3 + x^3\right)^{\frac{1}{3}} + 4k^2}}.$$

Hence it is evident, that when x is $= k$,

$$\frac{\sqrt{2k^2x - x^4} - \sqrt{kx^3}}{2x - \overline{7k^3 + x^3}|^{\frac{1}{3}}} \text{ is } = -\frac{10}{7}k.$$

COROLLARY. Seeing it amounts to the same thing, to take $x_{,}$ equal to x, and afterwards x equal to k; as to take, at first, $x_{,}$ and x each equal to k: it is obvious, that the quotients or fractions $\frac{N}{D}$ and $\frac{[x \perp N]}{[x \perp D]}$ become equal to each other, when x, in each, is taken equal to k.

EXAMPLE I. Supposing N and D equal to $\sqrt{2r^2x - x^4} - \sqrt{rx^3}$ and $2x - \overline{7r^3 + x^3}|^{\frac{1}{3}}$ respectively; we have $[x \perp N] = \frac{r^2 - 2x^3}{\sqrt{2r^2x - x^4}} - \frac{3}{2}r^{\frac{1}{2}}x^{\frac{1}{2}}$, and $[x \perp D] = 2 - \frac{x^2}{\overline{7r^3 + x^3}|^{\frac{2}{3}}}$. It appears then, that the value of $\frac{[x \perp N]}{[x \perp D]}$, when x is taken equal to r, is $= -\frac{10r}{7}$; and the same is the value of $\frac{N}{D}$, when x is so taken; which agrees with what is done in the preceding Example.

EXAMPLE II. If it be required to find the value of the quotient of $u\sqrt{1 - u_{,}^2} - u_{,}\sqrt{1 - u^2}$ divided by $u - u_{,}$, when $u_{,}$ is therein taken equal to u; we are to consider u as invariable: then, the special value of

$$\overline{\overline{u\sqrt{1 - u_{,}^2} - u_{,}\sqrt{1 - u^2}} - \overline{u\sqrt{1 - u_{,,}^2} - u_{,,}\sqrt{1 - u^2}}} \div$$

$\overline{u_{,} - u_{,,}}$ being $= -\frac{uu}{\sqrt{1 - u^2}} - \sqrt{1 - u^2}$ and $\overline{\overline{u - u_{,}} - \overline{u - u_{,,}}} \div \overline{u_{,} - u_{,,}}$ being $= -1$, we shall have $\frac{u^2}{\sqrt{1 - u^2}} + \sqrt{1 - u^2}$ $\left(= \frac{1}{\sqrt{1 - u^2}}\right)$ for the required value of $\overline{u\sqrt{1 - u_{,}^2} - u_{,}\sqrt{1 - u^2}} \div \overline{u - u_{,}}$.

Example III. Let s, c, and y be any functions of u; and let it be required to find the value of the quotient of

$$\overline{sy - y_{,} \times cu\sqrt{1-u_{,}^2} - cu_{,}\sqrt{1-u^2} + suu_{,} + s\sqrt{1-u^2}\times\sqrt{1-u_{,}^2}}$$

divided by $u - u_{,}$, when $u_{,}$ is therein taken equal to u.

Here we are to consider u, s, c, and y as invariable; and we may observe, that the proposed quotient is

$$= s \times \overline{y - uu_{,}y_{,} - y_{,}\sqrt{1-u^2}\times\sqrt{1-u_{,}^2}} \div \overline{u - u_{,}} -$$

$$cy_{,} \times \overline{u\sqrt{1-u_{,}^2} - u_{,}\sqrt{1-u^2}} \div \overline{u - u_{,}}.$$ Now the special value of $\overline{\overline{y - uu_{,}y_{,} - y_{,}\sqrt{1-u^2}\times\sqrt{1-u_{,}^2}} - \overline{y - uu_{,,}y_{,,} - y_{,,}\sqrt{1-u^2}\times\sqrt{1-u_{,,}^2}}}$

$\div \overline{u_{,} - u_{,,}}$ is $= - [u_{,} \perp y_{,}] \times \overline{uu_{,} + \sqrt{1-u^2} \times \sqrt{1-u_{,}^2}} -$

$uy_{,} + \dfrac{uy_{,}\sqrt{1-u^2}}{\sqrt{1-u_{,}^2}}$, and $\overline{\overline{u - u_{,}} - \overline{u - u_{,,}}} \div \overline{u_{,} - u_{,,}}$ is $= - 1$;

moreover, the value of $\overline{u\sqrt{1-u_{,}^2} - u_{,}\sqrt{1-u^2}} \div \overline{u - u_{,}}$

when $u_{,}$ is $= u$, is equal to $\dfrac{1}{\sqrt{1-u^2}}$, by the preceding Example: therefore, $[u_{,} \perp y_{,}]$ being $= [u \perp y]$ when $u_{,}$ is $= u$, and $y_{,}$ being then $= y$; $s[u \perp y] \div \dfrac{cy}{\sqrt{1-u^2}}$ is the required value of the quotient proposed.

Example IV. Suppose y to be any function of x: Then, $[x \perp y] - [x \mid y] = [x \perp y] - \dfrac{y - y_{,}}{x - x_{,}}$ being $=$ $\dfrac{\overline{x - x_{,}} \times [x \perp y] - \overline{y - y_{,}}}{x - x_{,}}$, we have $\dfrac{[x \perp y] - [x \mid y]}{x - x_{,}} = \dfrac{\overline{x - x_{,}} \times [x \perp y] - \overline{y - y_{,}}}{\overline{x - x_{,}}|^2}$,

where

where both numerator and denominator vanish, when $\underset{,}{x}$ is $= x$ ($\underset{,}{y}$ being then $= y$).

Therefore, by our rule, the quotients or fractions $\frac{[x \perp y] - [x \mid y]}{x - \underset{,}{x}}$ and $\frac{-[x \perp y] + [\underset{,}{x} \perp \underset{,}{y}]}{-2 \times \overline{x - \underset{,}{x}}}$ are equal to each other, when $\underset{,}{x}$, in each, is taken equal to x. But both numerator and denominator of the fraction $\frac{-[x \perp y] + [\underset{,}{x} \perp \underset{,}{y}]}{-2 \times \overline{x - \underset{,}{x}}}$ vanish, when $\underset{,}{x}$ is equal to x. Therefore, applying our rule a second time, it appears that the quotients or fractions $\frac{[x \perp y] - [x \mid y]}{x - \underset{,}{x}}$ and $\frac{[\underset{,}{x} \perp\!\!\perp \underset{,}{y}]}{2}$ are equal to each other, when $\underset{,}{x}$, in each, is taken equal to x.

Consequently the value of the quotient or fraction $\frac{[x \perp y] - [x \mid y]}{x - \underset{,}{x}}$, when $\underset{,}{x}$ is taken $= x$, is equal to $\frac{1}{2} \times [x \perp\!\!\perp y]$; which, it is plain, is the value of $\frac{[\underset{,}{x} \perp\!\!\perp \underset{,}{y}]}{2}$ when $\underset{,}{x}$ is taken as just now mentioned.

7.

y, z, &c. being any functions of x; and F and f being algebraic expressions composed, in any manner, of all, or any of, the quantities x, y, z, &c. and any invariable quantities:

If F be always $= f$,

$\underset{,}{F}$ will be $= \underset{,}{f}$,

$$F - \underset{,}{F} = f - \underset{,}{f},$$

$$\overline{F - \underset{,}{F}} \div \overline{x - \underset{,}{x}} = \overline{f - \underset{,}{f}} \div \overline{x - \underset{,}{x}},$$

and consequently $[x \perp F] = [x \perp f]$.

The

☞ The deducing this last equation from the first, (and likewise every similar operation,) I call *residual division*:—For the ready performing of which, the Corollaries to the first four Articles may serve as Rules.

EXAMPLE I. If ax be $= by^m$, we shall, by residual division, have

$$a = [x \perp by^m] = bmy^{m-1} \times [x \perp y];$$

$$\text{or } a \times [y \perp x] = [y \perp by^m] = bmy^{m-1};$$

$$\text{or } a \times [v \perp x] = [v \perp by^m] = bmy^{m-1} \times [v \perp y],$$

v being any function of x or y.

EXAMPLE II. If y be $= a + bx - cz$, we shall have

$$[x \perp y] = b - c \times [x \perp z],$$

$$\text{or} \quad 1 = b[y \perp x] - c[y \perp z],$$

$$\text{or} \quad [z \perp y] = b[z \perp x] - c.$$

EXAMPLE III. From the equation $ay^m = \overline{b^2 - x^2}|^n$, we get

$$amy^{m-1}[x \perp y] = -2nx \times \overline{b^2 - x^2}|^{n-1},$$

$$\text{or} \quad amy^{m-1} = -2nx[y \perp x] \times \overline{b^2 - x^2}|^{n-1};$$

$$\text{or } amy^{m-1} \times [v \perp y] = -2nx[v \perp x] \times \overline{b^2 - x^2}|^{n-1},$$

v being any function of x or y.

EXAMPLE IV. From the equation $x = ay^m z^n$, we find

$$1 = amz^n y^{m-1} \times [x \perp y] + any^m z^{n-1} \times [x \perp z].$$

EXAMPLE V. If x be $= \frac{y}{z} + \sqrt{y^2 + z^2} = yz^{-1} + \overline{y^2 + z^2}|^{\frac{1}{2}}$, we shall find, by our residual division,

$$1 = \frac{[x \perp y]}{z} - \frac{y[x \perp z]}{z^2} + \frac{y[x \perp y] + z[x \perp z]}{\overline{y^2 + z^2}|^{\frac{1}{2}}}.$$

EXAMPLE

EXAMPLE VI. $a^m + y^m$ being $= x^p \times \overline{a^n + x^n}|^q$, we get

$$my^{m-1}[x \perp y] = px^{p-1} \times \overline{a^n + x^n}|^q + nqx^{n+p-1} \times \overline{a^n + x^n}|^{q-1}.$$

EXAMPLE VII. From the equation $ax^2y + bxz^2 = cy^2z^2$, we get

$$2axy + ax^2[x \perp y] + bz^2 + 2bxz[x \perp z] = 2cyz^2[x \perp y] + 2cy^2z[x \perp z].$$

EXAMPLE VIII. $ax^m + bx^py^q + cy^n$ being $= 0$, we have

$$amx^{m-1} + bpy^qx^{p-1} + bqx^py^{q-1}[x \perp y] + cny^{n-1}[x \perp y] = 0.$$

EXAMPLE IX. Suppose $x = ay^2[x \perp y]$: then, by our division, we find

$$1 = 2ay[x \perp y]^2 + ay^2[x \perp\!\perp y].$$

EXAMPLE X. Let $axy + bx^2[x \perp y]$ be supposed $= 0$: then we shall have

$$ay + ax[x \perp y] + 2bx[x \perp y] + bx^2[x \perp\!\perp y] = 0.$$

EXAMPLE XI. $ax[x \perp z] + by[x \perp\!\perp y] + c[x \perp y]^2$ being $= 0$, we find

$$a[x \perp z] + ax[x \perp\!\perp z] + b[x \perp y][x \perp\!\perp y] + by[x \perp\!\perp\!\perp y] + 2c[x \perp y][x \perp\!\perp y] = 0.$$

EXAMPLE XII. Suppose $x^2 + axy + by^2 = 0$: then, by the division so often mentioned, we have

$$2x + ay + ax[x \perp y] + 2by[x \perp y] = 0:$$

And from hence, by a similar process, we get

$$2 + 2a[x \perp y] + ax[x \perp\!\perp y] + 2b[x \perp y]^2 + 2by[x \perp\!\perp y] = 0.$$

EXAMPLE XIII. If y be $= ax^m$; $[x \perp y]$ will be $= amx^{m-1}$, $[x \perp\!\perp y] = am . \overline{m-1} . x^{m-2}$, and $[x \perp\!\perp\!\perp y] = am . \overline{m-1} . \overline{m-2} . x^{m-3}$, &c. as appears by the method of division above-mentioned.

8.

8.

$\frac{z - \underset{.}{z}}{x - \underset{.}{x}}$ is manifestly $= \frac{z - \underset{.}{z}}{y - \underset{.}{y}} \div \frac{x - \underset{.}{x}}{y - \underset{.}{y}}$: it is obvious therefore, that, y and z being any functions of x,

$$[x \perp z] \text{ is } = \frac{[y \perp z]}{[y \perp x]}.$$

9.

$\frac{y - \underset{.}{y}}{x - \underset{.}{x}} \times \frac{x - \underset{.}{x}}{y - \underset{.}{y}}$ being $= 1$, $[x \perp y] \times [y \perp x]$ is evidently $= 1$,

$$\text{and } [x \perp y] = \frac{1}{[y \perp x]}.$$

From this last equation, we get, by residual division, $[x \underline{\perp} y] = [x \perp \frac{1}{[y \perp x]}]$, which (by what is shewn in the preceding Article) is $= [y \perp \frac{1}{[y \perp x]}] \times \frac{1}{[y \perp x]} = - \frac{[y \underline{\perp} x]}{[y \perp x]^3}$.

Moreover, from the equation $[x \underline{\perp} y] = - \frac{[y \underline{\perp} x]}{[y \perp x]^3}$, we find, by residual division, $[x \underline{\underline{\perp}} y] = - [x \perp \frac{[y \underline{\perp} x]}{[y \perp x]^3}] = - [y \perp \frac{[y \underline{\perp} x]}{[y \perp x]^3}] \times \frac{1}{[y \perp x]} = - \frac{[y \underline{\underline{\perp}} x]}{[y \perp x]^4} + \frac{3[y \underline{\perp} x]^2}{[y \perp x]^5}$, &c.

From what is done in this Chapter, it appears how the values of the quantities $[x \perp y]$, $[x \underline{\perp} y]$, &c. or $[y \perp x]$, $[y \underline{\perp} x]$, &c. may be obtained in terms of x and y, from an equation shewing the relation of the two quantities last mentioned, when no exponential quantity is concerned therein; which values we shall have frequent occasion to compute in the investigation of propositions by this Analysis. Rules for computing the values of the said quantities $[x \perp y]$, $[x \underline{\perp} y]$, &c. $[y \perp x]$, $[y \underline{\perp} x]$, &c. from equations containing exponentials, will be given in the next Chapter.

THE

THE

RESIDUAL ANALYSIS.

CHAP. III.

Of Exponentials and Logarithms.

AS one variable quantity may be denoted by the algebraic expreſſion x^n, or the like, where the Root only is variable whilſt the Exponent remains invariable; ſo may another variable quantity be denoted by n^x, or the like, where the Exponent only is variable whilſt the Root remains invariable. Therefore, having ſhewn how $x^n - \underset{,}{x}^n$, may be divided by $x - \underset{,}{x}$, it will now be proper to ſhew how $n^x - n^{\underset{,}{x}}$ may alſo be divided by the ſame diviſor $(x - \underset{,}{x})$.—In doing which, we ſhall firſt aſſign the value of n^x in a certain ſeries of terms of n and x, wherein the exponents of the ſeveral powers of theſe quantities ſhall be invariable: by help of which ſeries we ſhall be enabled readily to obtain the deſired quotient of $n^x - n^{\underset{,}{x}}$ divided by $x - \underset{,}{x}$.

We shall then shew how, by means of that quotient, the value of n^x may be assigned in another series of terms of n and x, in which the exponents shall be invariable; and likewise how, from the equation $n^x = v$, the value of x may be assigned in a series of terms of n and v, with invariable exponents.—From which series many useful conclusions relating to Exponentials and Logarithms will be easily deduced.

I.

Assuming $x^{\frac{m}{r}} = 1 + b.\overline{x-1} + c.\overline{x-1}|^2 + d.\overline{x-1}|^3$ &c. *it is proposed to find, in terms of* m *and* r, *the coefficients* b, c, d, *&c.*

Writing (in the assumed equation) y instead of x, we have

$$y^{\frac{m}{r}} = 1 + b.\overline{y-1} + c.\overline{y-1}|^2 + d.\overline{y-1}|^3 \text{ \&c.}$$

and by subtraction,

$$x^{\frac{m}{r}} - y^{\frac{m}{r}} = b.\overline{x-y} + c.\overline{\overline{x-1}|^2 - \overline{y-1}|^2} + d.\overline{\overline{x-1}|^3 - \overline{y-1}|^3} \text{ \&c.}$$

If, now, we divide by the residual $x-y$, we shall get

$$x^{\frac{m}{r}-1} \times \frac{1 + \frac{y}{x} \cdots + \overline{\tfrac{y}{x}}|^2 \cdots (m)}{1 + \overline{\tfrac{y}{x}}|^{\frac{m}{r}} + \overline{\tfrac{y}{x}}|^{\frac{2m}{r}} \cdots (r)} =$$

$$b + c.\overline{\overline{x-1} + \overline{y-1}} + d.\overline{\overline{x-1}|^2 + \overline{x-1} \times \overline{y-1} + \overline{y-1}|^2}$$

&c. which equation must hold true let y be what it will: From whence, by taking y equal to x, we find

$$\frac{m}{r} \times x^{\frac{m}{r}-1} = b + 2c.\overline{x-1} + 3d.\overline{x-1}|^2 \text{ \&c.}$$

Conse-

Conſequently, multiplying one ſide by x and the other by $1+\overline{x-1}$, we have $\frac{m}{r}\times x^{\frac{m}{r}}$, or its equal

$$\frac{m}{r}+\frac{m}{r}b.\overline{x-1}+\frac{m}{r}c.\overline{x-1}|^{2}+\frac{m}{r}d.\overline{x-1}|^{3}\ \text{\&c.}$$

$$=b+\left.\begin{matrix}2c\\ b\end{matrix}\right\}\overline{x-1}+\left.\begin{matrix}3d\\ 2c\end{matrix}\right\}\overline{x-1}|^{2}+\left.\begin{matrix}4e\\ 3d\end{matrix}\right\}\overline{x-1}|^{3}\ \text{\&c.}$$

Hence, by equating the homologous terms, it appears, that b is $=\frac{m}{r}$, $c=\frac{m-r}{2r}.b$, $d=\frac{m-2r}{3r}.c$, $e=\frac{m-3r}{4r}.d$, &c.

In like manner, if we aſſume

$$\frac{1}{x^{\frac{m}{r}}}=1+b.\overline{x-1}+c.\overline{x-1}|^{2}+d.\overline{x-1}|^{3}\ \text{\&c.}$$

we ſhall find $b=-\frac{m}{r}$, $c=-\frac{m+r}{2r}.b$, $d=-\frac{m+2r}{3r}.c$, &c.

Conſequently, $\frac{1}{x^{\frac{m}{r}}}$ being $=x^{-\frac{m}{r}}$, it is manifeſt, that, whether the exponent $\frac{m}{r}$ be poſitive or negative, $x^{\frac{m}{r}}$ is $=$ $1+\frac{m}{r}.\overline{x-1}+\frac{m}{r}.\frac{m-r}{2r}.\overline{x-1}|^{2}+\frac{m}{r}.\frac{m-r}{2r}.\frac{m-2r}{3r}.\overline{x-1}|^{3}$ &c.

Hence, by writing n inſtead of x, and x inſtead of $\frac{m}{r}$, we have $n^{x}=1+x.\overline{n-1}+x.\frac{x-1}{2}.\overline{n-1}|^{2}+x.\frac{x-1}{2}.\frac{x-2}{3}.\overline{n-1}|^{3}$ &c. x being any number whatever.

And, by ſubſtituting, in this laſt equation, $\frac{a+z}{a}$ inſtead of n, it appears that $\overline{a+z}|^{x}$ is $=a^{x}+x.a^{x-1}z+x.\frac{x-1}{2}.a^{x-2}z^{2}$ $+x.\frac{x-1}{2}.\frac{x-2}{3}.a^{x-3}z^{3}$ &c. which is the Binomial Theorem *.

2. The

* The ſame concluſion, by referring to what is proved in the preceding Chapter, may be obtained in the following brief manner; but I preſumed, that,

2.

The expreſſion $\frac{n^x - n^{x_{,}}}{x - x_{,}}$ being $= n^{x_{,}} \times \frac{n^{x - x_{,}} - 1}{x - x_{,}}$, which by the preceding Article is

$$= n^{x_{,}} \times \overline{n-1} + \frac{x - x_{,} - 1}{2} \cdot \overline{n-1}|^2 + \frac{x - x_{,} - 1}{2} \cdot \frac{x - x_{,} - 2}{3} \cdot \overline{n-1}|^3$$

&c. it is obvious, that, when $x_{,}$ is equal to x, the value of the quotient of $n^x - n^{x_{,}}$ divided by $x - x_{,}$ is

$$= n^x \times \overline{\overline{n-1} - \frac{\overline{n-1}|^2}{2} + \frac{\overline{n-1}|^3}{3} - \frac{\overline{n-1}|^4}{4}\ \&c.} = g \times n^x,$$

g being put for the ſeries $\overline{n-1} - \frac{\overline{n-1}|^2}{2} + \frac{\overline{n-1}|^3}{3}$ — &c. which converges when n is a poſitive number leſs than 2.

Writing $\frac{1}{n}$ inſtead of n, we get

$$\text{ſpec. val. of } \overline{\frac{1}{n^x} - \frac{1}{n^{x_{,}}}} \div \overline{x - x_{,}} = -\frac{1}{n^{x+x_{,}}} \times \text{ſpec. val. of } \frac{n^x - n^{x_{,}}}{x - x_{,}}$$

$$= \frac{1}{n^x} \times \overline{\frac{1-n}{n} - \frac{1}{2} \cdot \overline{\frac{1-n}{n}}|^2 + \frac{1}{3} \cdot \overline{\frac{1-n}{n}}|^3 - \&c.}$$

---

that, upon the firſt application of what is there taught, it would not be amiſs to give a more explicit proceſs.

The brief inveſtigation is this.

Aſſuming $\overline{1+v}|^x = 1 + bv + cv^2 + dv^3$ &c.

we get, by reſidual diviſion, ($v - v_{,}$ being the diviſor,)

$$x \times \overline{1+v}|^{x-1} = b + 2cv + 3dv^2\ \&c.$$

Conſequently, multiplying by $1 + v$, we have

$$x \times \overline{1+v}|^x, \text{ or its equal } x + xbv + xcv^2 + xdv^3\ \&c.$$

$$= b + \left.\begin{matrix}2c\\b\end{matrix}\right\} v + \left.\begin{matrix}3d\\2c\end{matrix}\right\} v^2 + \left.\begin{matrix}4e\\3d\end{matrix}\right\} v^3\ \&c.$$

From whence, by equating the homologous terms, b, c, d, &c. will be found equal to x, $\frac{x \cdot \overline{x-1}}{2}$, $\frac{x \cdot \overline{x-1} \cdot \overline{x-2}}{2 \cdot 3}$, &c. reſpectively: And then the value of $\overline{a+z}|^x$ may be obtained by ſubſtituting $\frac{z}{a}$ inſtead of v.

From

From whence we find, that the *spec. val.* of $\overline{n^{x} - n^{x_{,}}} \div \overline{x - x_{,}}$ is also $= n^{x} \times \overline{\frac{n-1}{n} + \frac{1}{2} \cdot \overline{\frac{n-1}{n}}\Big|^{2} + \frac{1}{3} \cdot \overline{\frac{n-1}{n}}\Big|^{3} \text{ \&c.}} = g \times n^{x}$, g being now put for the series $\frac{n-1}{n} + \frac{1}{2} \cdot \overline{\frac{n-1}{n}}\Big|^{2}$ &c. which converges when n is any positive number greater than $\frac{1}{2}$.

3.

Assuming $n^{x} = 1 + Ax + Bx^{2} + Cx^{3}$ &c. *it is proposed to find, in terms of* n, *the coefficients* A, B, C, &c.

$$n^{x_{,}} \text{ being } = 1 + Ax_{,} + Bx_{,}^{2} + Cx_{,}^{3} \text{ \&c.}$$

we have, by subtraction,

$$n^{x} - n^{x_{,}} = A \cdot \overline{x - x_{,}} + B \cdot \overline{x^{2} - x_{,}^{2}} + C \cdot \overline{x^{3} - x_{,}^{3}} \text{ \&c.}$$

and from hence, by division,

$$\frac{n^{x} - n^{x_{,}}}{x - x_{,}} = A + B \cdot \overline{x + x_{,}} + C \cdot \overline{x^{2} + xx_{,} + x_{,}^{2}} \text{ \&c.}$$

Now, the quotient of $n^{x} - n^{x_{,}}$ divided by $x - x_{,}$ being, by the last Article, equal to $g \times n^{x}$ when $x_{,}$ is $= x$, it follows, that

$$g \times n^{x}, \text{ or its equal } g + gAx + gBx^{2} + gCx^{3} \text{ \&c.}$$
$$\text{is} = A + 2Bx + 3Cx^{2} + 4Dx^{3} \text{ \&c.}$$

From whence, by comparing the homologous terms, A is found $= g$, $B = \frac{gA}{2}$, $C = \frac{gB}{3}$, $D = \frac{gC}{4}$, &c.

Therefore n^{x} is $= 1 + gx + \frac{g^{2}x^{2}}{2} + \frac{g^{3}x^{3}}{2 \cdot 3} + \frac{g^{4}x^{4}}{2 \cdot 3 \cdot 4}$ &c.

4.

From the equation $n^{x} = v$, *it is now proposed to find* x *in terms of* n *and* v.

If

If x be assumed $= m\times\overline{v-1}+A.\overline{v-1}|^{2}+B.\overline{v-1}|^{3}$ &c. and $n^{x} = v$, (m, A, B, &c. being supposed independent of v,) we shall have

$$x_{,} = m\times\overline{v_{,}-1}+A.\overline{v_{,}-1}|^{2}+B.\overline{v_{,}-1}|^{3} \text{ \&c.}$$

and, by subtraction,

$$x - x_{,} = m\times\overline{v-v_{,}} + A.\overline{\overline{v-1}|^{2}-\overline{v_{,}-1}|^{2}}+B.\overline{\overline{v-1}|^{3}-\overline{v_{,}-1}|^{3}} \text{ \&c.}$$

From which last equation, by dividing one side by $n^{x} - n^{x_{,}}$ and the other by $v - v_{,}$ ($=\overline{v-1}-\overline{v_{,}-1}$), we get

$$\frac{x-x_{,}}{n^{x}-n^{x_{,}}} = m\times 1 + A.\overline{v-1+v_{,}-1}+B.\overline{\overline{v-1}|^{2}+\overline{v-1}\times\overline{v_{,}-1}+\overline{v_{,}-1}|^{2}} \text{ \&c.}$$

Hence, by taking $x_{,}$ equal to x, and $v_{,}$ equal to v, we have

$$\frac{1}{g\times n^{x}} = m\times 1+2A.\overline{v-1}+3B.\overline{v-1}|^{2}+4C.\overline{v-1}|^{3} \text{ \&c.}$$

$g\times n^{x}$, by Art. 2, being the value of the quotient of $n^{x}-n^{x_{,}}$ divided by $x-x_{,}$, when $x_{,}$ is therein taken equal to x.

But $\frac{1}{n^{x}}$ is $=\frac{1}{v}=\frac{1}{1+\overline{v-1}}$, which, by division, is found equal to $1-\overline{v-1}+\overline{v-1}|^{2}-\overline{v-1}|^{3}+\overline{v-1}|^{4}-$&c.

It is evident therefore, that

$$\frac{1}{g}\times\overline{1-\overline{v-1}+\overline{v-1}|^{2}-\overline{v-1}|^{3}+\overline{v-1}|^{4}-\text{\&c.}}$$

is $= m\times 1+2A.\overline{v-1}+3B.\overline{v-1}|^{2}+4C.\overline{v-1}|^{3}+5D.\overline{v-1}|^{4}$ &c.

From whence, by comparing the homologous terms, we find $m=\frac{1}{g}$, $A=-\frac{1}{2}$, $B=\frac{1}{3}$, $C=-\frac{1}{4}$, $D=\frac{1}{5}$, &c. and consequently $x=\frac{1}{g}\times\overline{\overline{v-1}-\frac{\overline{v-1}|^{2}}{2}+\frac{\overline{v-1}|^{3}}{3}-\frac{\overline{v-1}|^{4}}{4}+\text{\&c.}}$

5. By

5.

By putting $1 + u$ equal to $n^x = v$, we have, by Article 3,

$$1 + u = 1 + gx + \frac{g^2x^2}{2} + \frac{g^3x^3}{2\cdot3} + \frac{g^4x^4}{2\cdot3\cdot4}\ \&c.$$

and from the equation

$$x = \frac{1}{g}\times\overline{v - 1 - \frac{\overline{v-1}|^2}{2} + \frac{\overline{v-1}|^3}{3} - \&c.}$$

$$x = \frac{1}{g}\times\overline{u - \frac{u^2}{2} + \frac{u^3}{3} - \frac{u^4}{4} + \&c.}$$

where g is $= \overline{n-1} - \frac{\overline{n-1}|^2}{2} + \frac{\overline{n-1}|^3}{3} - \frac{\overline{n-1}|^4}{4} + \&c.$

or $\frac{n-1}{n} + \frac{1}{2}\cdot\overline{\frac{n-1}{n}}\Big|^2 + \frac{1}{3}\cdot\overline{\frac{n-1}{n}}\Big|^3\ \&c.$

COROLLARY I. *Logarithms* being an artificial Set of numbers corresponding to another Set of numbers in such a manner, that the sum of the logarithms of any two numbers is equal to the logarithm of the product of those two numbers *; and the sum of the exponents of any two powers of any given quantity being equal to the exponent of that power (of the same given quantity) which is produced by multiplying those two powers together; it is obvious that the exponents of the powers of any given quantity are logarithms of the values of those powers respectively: For instance, a, b, c, d, &c. are logarithms of n^a, n^b, n^c, n^d, &c. respectively.

It follows therefore, from what is said above, that

$\frac{1}{g}\times\overline{v-1-\frac{\overline{v-1}|^2}{2}+\frac{\overline{v-1}|^3}{3}-\&c.}$ $\frac{1}{g}\times\overline{u-\frac{u^2}{2}+\frac{u^3}{3}-\&c.}$ and $\frac{1}{g}\times\overline{w-\frac{w^2}{2}+\frac{w^3}{3}-\&c.}$ are logarithms of v, $1 + u$,

* From this property of logarithms, it plainly follows, that, b, c, p, and N being any numbers whatever, the Log. of b − the Log. of c is = the Log. of $\frac{b}{c}$, and p × Log. of N = Log. of N^p.

and $1 + w$ respectively, in the same Set, g being of any determinate value whatever.

From the equation, Log. of $\overline{1+u} = \frac{1}{g} \times \overline{u - \frac{u^2}{2} + \frac{u^3}{3} - \&c.}$

we, by writing $\frac{1}{u}$ instead of u, have Log. of $1 + \frac{1}{u}$ (= Log. of $\overline{1+u}$ — Log. of u) $= \frac{1}{g} \times \overline{u^{-1} - \frac{u^{-2}}{2} + \frac{u^{-3}}{3} - \&c.}$ and consequently

Log. of $\overline{1+u}$ = Log. of $u + \frac{1}{g} \times \overline{u^{-1} - \frac{u^{-2}}{2} + \frac{u^{-3}}{3} - \&c.}$

which last series converges when the series in the former value of the Log. of $\overline{1+u}$ does not converge.

So from the equat. Log. of $v = \frac{1}{g} \times \overline{\overline{v-1} - \frac{\overline{v-1}|^2}{2} + \frac{\overline{v-1}|^3}{3} - \&c.}$ (where v must be a positive number less than 2,) we, by substituting $\frac{1}{v}$ instead of v, find

$$\text{Log. of } v = \frac{1}{g} \times \overline{\frac{v-1}{v} + \frac{1}{2} \cdot \overline{\frac{v-1}{v}}\Big|^2 + \frac{1}{3} \cdot \overline{\frac{v-1}{v}}\Big|^3 \ \&c.}$$

where v may be any positive number greater than $\frac{1}{2}$.

Now, as in these logarithmic expressions the value of $\frac{1}{g}$ may be taken at pleasure, so that in any one Set it be always the same; it is plain, that, to any Set of numbers, we may adapt as many Sets of logarithms as we please, by taking $\frac{1}{g}$ of different values in different Sets; and the logarithm of any number, in any one Set of logarithms, being to the logarithm of the same number, in any other Set of logarithms, as the value of the factor $\frac{1}{g}$ in the former Set is to the value of $\frac{1}{g}$ in the latter, that factor is called the *Modulus* of the Set.

SCHOLIUM. *Lord* NAPIER, the Inventor of Logarithms, first gave a Set, whereof the modulus is Unity:—In which Set the logarithm

logarithm of 10 is 2.302585 * :—Of this Set are the logarithms now called *Napier's logarithms* †. And he afterwards, with the assistance of Mr. BRIGGS, (then *Savilian* Professor of Geometry, at *Oxford*,) undertook to compute a second Set of logarithms, wherein the logarithm of 10 is 1; (which Set he found would be more convenient for trigonometrical calculations than his former;) but, Lord NAPIER dying before they had finished their design, the work was completed by Mr. BRIGGS, and the logarithms of this second Set (which are the common tabular logarithms) are called *Briggs's logarithms*. The modulus of this second Set is $\frac{1}{2.302585} = 0.4342944$: for, by what is said above, the logarithm of 10, which in this Set is 1, is $= \frac{1}{g} \times 2.302585$; and, consequently, $\frac{1}{g} = \frac{1}{2.302585}$.

COROLLARY II. From what is said above, it plainly appears, that if, in any Set of logarithms whose modulus is $\frac{1}{g}$, x be the logarithm of any number whatsoever, that number will be equal to

$$1 + gx + \frac{g^2x^2}{2} + \frac{g^3x^3}{2 \cdot 3} + \frac{g^4x^4}{2 \cdot 3 \cdot 4} \text{ \&c.} = n^x.$$

COROLLARY III. By taking, in the last Corollary, x equal to $\frac{1}{g}$, it appears that, in any Set of logarithms, the modulus is always the logarithm of $1 + 1 + \frac{1}{2} + \frac{1}{2 \cdot 3} + \frac{1}{2 \cdot 3 \cdot 4}$ &c. $(= 2.7182818)$; or, which is the same thing, of the ratio of the sum of that Series to Unity: which ratio is therefore, by Mr. COTES, called the *modular ratio*.

And, $-x$ being the logarithm of the reciprocal of any number whose logarithm is x; it appears by taking, in the same Corollary, x equal to $-\frac{1}{g}$, that the modular ratio is that of Unity to the sum of the series $1 - 1 + \frac{1}{2} - \frac{1}{2 \cdot 3} + \frac{1}{2 \cdot 3 \cdot 4} -$ &c. $(= 0.3678794)$.

* See Scholium to Article 6. † These are also called *Hyperbolic logarithms*, for a reason that will appear in a subsequent Chapter.

Corollary IV. It is evident then, that, in *Napier's* Set, (whereof the modulus is 1,) 2.7182818 is the number whose logarithm is 1; and that 0.3678794 is the number whose logarithm is -1: Also, that, n being $= 2.7182818$,

$$n^x \text{ is } = 1 + x + \frac{x^2}{2} + \frac{x^3}{2 \cdot 3} + \frac{x^4}{2 \cdot 3 \cdot 4} \text{ \&c.}$$

6.

Napier's Log. of $\overline{1+x}$ being, by Art. 5, $= x - \frac{x^2}{2} + \frac{x^3}{3} - \frac{x^4}{4} + \text{\&c}$ it appears, by writing $-x$ instead of x, that *Napier's Log.* of $\overline{1-x}$ is $= -x - \frac{x^2}{2} - \frac{x^3}{3} - \frac{x^4}{4}$ &c. and, by subtraction, that *Nap Log.* of $\overline{1+x}$ — *Nap. Log.* of $\overline{1-x}$ (= *Nap. Log.* of $\frac{1+x}{1-x}$) is equal to $2 \times \overline{x + \frac{x^3}{3} + \frac{x^5}{5} + \frac{x^7}{7} \text{ \&c.}}$

By this theorem the logarithms of small numbers may be easily computed.

Example I. *To find Napier's logarithm of 2.*

Supposing $\frac{1+x}{1-x} = 2$, x will from thence be found $= \frac{1}{3}$, and therefore *Nap. Log.* of 2 is $= 2 \times \overline{\frac{1}{1 \cdot 3} + \frac{1}{3 \cdot 3^3} + \frac{1}{5 \cdot 3^5} \text{ \&c.}}$ $= 0.69314718$.

Example II. *To find Napier's logarithm of* $\frac{5}{4}$.

From the equation $\frac{1+x}{1-x} = \frac{5}{4}$ we find $x = \frac{1}{9}$; and, consequently, *Nap. Log.* of $\frac{5}{4} = 2 \times \overline{\frac{1}{1 \cdot 9} + \frac{1}{3 \cdot 9^3} + \frac{1}{5 \cdot 9^5} \text{ \&c.}}$ $= 0.22314355$.

Scholium. To find the logarithm of a great number, by the same theorem, we must find the logarithms of such small numbers as, being multiplied together, shall produce that great number; and then, the sum of the logarithms of any numbers being equal to the logarithm of the product of those numbers, we shall, by adding those logarithms together, find the logarithm required.

Example.

EXAMPLE. *To find Napier's logarithm of* 10.

It is obvious, that 10 is $= 8\times\frac{5}{4} = 2\times 2\times 2\times\frac{5}{4}$; therefore, by what is just now said, *Napier's logarithm* of 2 being computed, and likewise his logarithm of $\frac{5}{4}$, the sum of this last logarithm added to three times the former will be (=2.30258509) the required logarithm of 10.

7.

Other theorems, more convenient for computing the logarithms of some certain numbers, may be deduced from the above; of which the following are examples.

EXAMPLE I. *To find the logarithm of* n, *having the logarithms of* n − 1 *and* n + 1 *given.*

Supposing $\frac{n^2}{\overline{n-1}\times\overline{n+1}}$, or its equal $\frac{n^2}{n^2-1} = \frac{1+x}{1-x}$, we from thence find $x = \frac{1}{2n^2-1}$. Therefore, by the theorem in the last Article, *Nap. Log.* of $\frac{n^2}{n^2-1}$ (= 2 × *Nap. Log.* of n − *Nap. Log.* of $\overline{n-1}$ − *Nap. Log.* of $\overline{n+1}$) is =

$$2\times\overline{\frac{1}{2n^2-1}+\frac{1}{3\cdot\overline{2n^2-1}|^3}+\frac{1}{5\cdot\overline{2n^2-1}|^5}}\ \&c.$$

and, consequently,

$$\textit{Nap. Log. of } n = \frac{\textit{Nap. Log. of } \overline{n-1} + \textit{Nap. Log. of } \overline{n+1}}{2} + \frac{1}{2n^2-1} + \frac{1}{3\cdot\overline{2n^2-1}|^3} + \frac{1}{5\cdot\overline{2n^2-1}|^5}\ \&c.$$

This theorem is very convenient for computing the logarithm of any large prime number.

EXAMPLE II. *To find the logarithm of a large number* n, *having the logarithms of* n − x *and* n + x *given.*

 The

The Log. of $1 - x$ being $= -$ Mod. $\times \overline{x + \frac{x^2}{2} + \frac{x^3}{3}}$ &c. the Log. of $\frac{1}{1-x}$ is $=$ Mod. $\times \overline{x + \frac{x^2}{2} + \frac{x^3}{3}}$ &c. which (the Log. of $\frac{1+x}{1-x}$ being $= 2 \times$ Mod. $\times \overline{x + \frac{x^3}{3} + \frac{x^5}{5}}$ &c.) is $= \frac{1}{2} \times$ the Log. of $\overline{\frac{1+x}{1-x}} \times \frac{x + \frac{x^2}{2} + \frac{x^3}{3} \text{ \&c.}}{x + \frac{x^3}{3} + \frac{x^5}{5} \text{ \&c.}} = \frac{1}{2} \times$ Log. of $\overline{\frac{1+x}{1-x}} \times \overline{1 + \frac{x}{2} + \frac{x^3}{12} + \frac{7x^5}{180}}$ &c. Hence, by substituting $\frac{x}{n}$ instead of x, we have Log. of $\frac{n}{n-x}$ (= Log. of n — Log. of $\overline{n-x}$) $= \frac{1}{2} \times$ Log. of $\overline{\frac{n+x}{n-x}} \times \overline{1 + \frac{x}{2n} + \frac{x^3}{12n^3} + \frac{x^5}{180n^5}}$ &c. Therefore, the Log. of $\frac{n+x}{n-x}$ being $=$ Log. of $\overline{n+x}$ — Log. of $\overline{n-x}$, the Log. of n is $= \frac{1}{2} \times$ Log. of $\overline{n+x} + \frac{1}{2} \times$ Log. of $\overline{n-x} +$ $\frac{\text{Log. of } \overline{n+x} - \text{Log. of } \overline{n-x}}{2} \times \overline{\frac{x}{2n} + \frac{x^3}{12n^3} + \frac{x^5}{180n^5}}$ &c.

This last theorem is very useful in enlarging a table of logarithms.

8.

It may sometimes be of use to observe, that, when v is a very large number, the Log. of $1 + \frac{1}{v}$ will be $= \frac{1}{v}$ nearly, the value of the series $\frac{v^{-2}}{2} - \frac{v^{-3}}{3} + \frac{v^{-4}}{4}$ — &c. (See Art. 5. Cor. 1.) being then so very small that it may be neglected. Therefore, v being as just now mentioned, if $\overline{1 + \frac{1}{v}}\big|^{u}$ be $=$ N, the Log. of N ($= u \times$ Log. of $\overline{1 + \frac{1}{v}}$) will be nearly equal to $\frac{u}{v}$.

9. By

9.

By Art. 2. $[x \perp n^x]$ is $= gn^x$, g being (by that Article, and Cor. 1. Art. 5.) $=$ *Nap. Log.* of n.

But in *Napier*'s Set, 2.7182818 is the number whose logarithm is 1, as appears by Cor. 4. Art. 5.

Therefore if n be $= 2.7182818$ (*i. e.* if the ratio of n to 1 be the modular ratio) g will here be $= 1$; and, in that case, $[x \perp n^x]$ is $= n^x$; or, v being $= n^x$, $[x \perp v]$ is $= v$.

Consequently $[x \perp v] \times [v \perp x]$ being $= 1$, (by Chap. 2. Art. 9.)

$$[v \perp x] \text{ is } = \frac{1}{v}, \text{ where } x \text{ is } \textit{Nap. Log.} \text{ of } v.$$

Moreover, by Chap. 2. Art. 8. $[v \perp x]$ is equal to $\frac{[u \perp x]}{[u \perp v]}$: this last expression must therefore be $= \frac{1}{v}$; and, consequently,

$$[u \perp x] = \frac{[u \perp v]}{v},$$

u being any function of v, or x.

☞ For brevity sake, we shall, in future, put $L:x$, $L:\overline{a+bx}$, $L:\overline{x+\sqrt{1+x^2}}|^p$, &c. to denote *Nap. Log.* of x, *Nap. Log.* of $\overline{a+bx}$, *Nap. Log.* of $\overline{x+\sqrt{1+x^2}}|^p$, &c. respectively.

EXAMPLE I. *Suppose* $v = 1 + z$: then we have

$$[u \perp L:\overline{1+z}] = \frac{[u \perp z]}{1+z}.$$

EXAMPLE II. *Supposing* $v = a + bz^m$, we have

$$[u \perp L:\overline{a+bz^m}] = \frac{[u \perp \overline{a+bz^m}]}{a+bz^m} = \frac{bmz^{m-1}[u \perp z]}{a+bz^m}.$$

EXAMPLE III. *Let* v *be supposed* $= a + z + \sqrt{2az+z^2}$.

Then we have $[u \perp L:\overline{a+z+\sqrt{2az+z^2}}] = \frac{[u \perp z]}{\sqrt{2az+z^2}}$,

[$u \perp$

$[u \perp \overline{a+z+\sqrt{2az+z^2}}]$ being $= \frac{a+z+\sqrt{2az+z^2}}{\sqrt{2az+z^2}} \times [u \perp z]$.

Example IV. $[u \perp L : \overline{z^n + \sqrt{a^2+z^{2n}}}]$ is $= \frac{nz^{n-1}[u \perp z]}{\sqrt{a^2+z^{2n}}}$,

$[u \perp \overline{z^n + \sqrt{a^2+z^{2n}}}]$ being $= \overline{z^n + \sqrt{a^2+z^{2n}}} \times \frac{nz^{n-1}[u \perp z]}{\sqrt{a^2+z^{2n}}}$.

Example V. *v being supposed* $= \frac{a+z}{a-z}$, we have

$$[u \perp L : \frac{a+z}{a-z}] = \frac{2a[u \perp z]}{\overline{a-z}|^2} \div \frac{a+z}{a-z} = \frac{2a[u \perp z]}{a^2-z^2}.$$

Example VI. *Taking* $v = \frac{a-\sqrt{a^2+z^2}}{a+\sqrt{a^2+z^2}}$, we find

$$[u \perp L : \frac{a-\sqrt{a^2+z^2}}{a+\sqrt{a^2+z^2}}] = \frac{2a[u \perp z]}{z\sqrt{a^2+z^2}}.$$

10.

By the laſt Article we have $[x \perp L : r^v] = \frac{[x \perp r^v]}{r^v}$. Now $L : r^v$ being $= v \times L : r$, $[x \perp L : r^v]$ is $= [x \perp v \times L : r] = [x \perp v] \times L : r$, v being any function of the variable quantity x, and r invariable: Therefore $\frac{[x \perp r^v]}{r^v}$ is then $= [x \perp v] \times L : r$; and, conſequently, $[x \perp r^v] = r^v \times [x \perp v] \times L : r$.

But if both r and v be functions of x, $[x \perp v \times L : r]$ $(= [x \perp L : r^v] = \frac{[x \perp r^v]}{r^v})$ will be $= [x \perp v] \times L : r + v \times \frac{[x \perp r]}{r}$: Conſequently $[x \perp r^v]$ will then be $= r^v \times [x \perp v] \times L : r + r^{v-1} v [x \perp r]$.

By proceeding much in the ſame manner, may the values of $[x \perp \overline{r^v}|^w]$ and $[x \perp \overline{r}|^{v^w}]$, &c. be found in terms of r, v, w, x, $[x \perp r]$, $[x \perp v]$, $[x \perp w]$, &c.

Exam-

EXAMPLE I. N and n being invariable, $[x \perp N^{nx}]$ is $= nN^{nx} \times L : N$.

EXAMPLE II. If a, b, and n be invariable; $[x \perp \overline{a+bx}|^{nx}]$ will be $= n \times \overline{a+bx}|^{nx} \times L : \overline{a+bx} + bnx \times \overline{a+bx}|^{nx-1}$.

EXAMPLE III. If (A, N, and a, b, c, &c. being invariable)

$$AN^{x} \text{ be } = \overline{y+a} \,.\, \overline{y+b} \,.\, \overline{y+c}\ (n).$$

we, by refidual divifion, fhall have

$$AN^{x} \times [y \perp x] \times L : N = \overline{y+b} \,.\, \overline{y+c} \,.\, \overline{y+d}\ (n-1) + \overline{y+a} \,.\, \overline{y+c} \,.\, \overline{y+d}\ (n-1) + \overline{y+a} \,.\, \overline{y+b} \,.\, \overline{y+d}\ (n-1) + (n).$$

II.

By the binomial theorem, inveftigated in Article 1, we have

$$\overline{1+z}|^{v} = 1 + vz + \frac{v \,.\, v-1}{2} \,.\, z^{2} + \frac{v \,.\, v-1 \,.\, v-2}{2 \,.\, 3} \,.\, z^{3} \text{ \&c.}$$

from whence, by refidual divifion, ($v - \acute{v}$ being the divifor, and the value of z independent of v,) we get

$$\overline{1+z}|^{v} \times L : \overline{1+z} = z + \left.\begin{matrix} v-1 \\ v \end{matrix}\right\} \frac{z^{2}}{2} + \left.\begin{matrix} v-1 \,.\, v-2 \\ v \,.\, v-2 \\ v \,.\, v-1 \end{matrix}\right\} \frac{z^{3}}{2 \,.\, 3} \text{ \&c.}$$

Hence, fuppofing $v = 0$, or any pofitive integer, we find

$$L : \overline{1+z} = \frac{P}{\overline{1+z}|^{v}} + \frac{1 \,.\, 1 \,.\, 2 \,.\, 3\,(v+1)}{\overline{1+z}|^{v}} \times \overline{\frac{z^{v+1}}{1 \,.\, 2 \,.\, 3\,(v+1)} - \frac{z^{v+2}}{2 \,.\, 3 \,.\, 4\,(v+1)} + \frac{z^{v+3}}{3 \,.\, 4 \,.\, 5\,(v+1)} - \frac{z^{v+4}}{4 \,.\, 5 \,.\, 6\,(v+1)}} + \text{\&c.}$$

P being put for $z + \left.\begin{matrix} v-1 \\ v \end{matrix}\right\} \frac{z^{2}}{2} + \left.\begin{matrix} v-1 \,.\, v-2 \\ v \,.\, v-2 \\ v \,.\, v-1 \end{matrix}\right\} \frac{z^{3}}{2 \,.\, 3}\ (v)$.

12. Again,

12.

Again, by the binomial theorem we have $\overline{z+d}|^{x+1} = z^{x+1} + \overline{x+1} \cdot dz^x + \frac{\overline{x+1} \cdot x}{2} d^2 z^{x-1} + \frac{\overline{x+1} \cdot x \cdot \overline{x-1}}{2 \cdot 3} d^3 z^{x-2}$ &c.

Therefore

$$z^x \text{ is } = \frac{\overline{z+d}|^{x+1} - z^{x+1}}{d \cdot \overline{x+1}} - \frac{x}{2} dz^{x-1} - \frac{x \cdot \overline{x-1}}{2 \cdot 3} d^2 z^{x-2} \text{ \&c.}$$

Hence, expounding z by a, $a+d$, $a+2d$, &c. successively, we have

$$a^x = \frac{\overline{a+d}|^{x+1} - a^{x+1}}{d \cdot \overline{x+1}} - \frac{x}{2} d \cdot a^{x-1} - \frac{x \cdot \overline{x-1}}{2 \cdot 3} d^2 \cdot a^{x-2} \text{ \&c.}$$

$$\overline{a+d}|^x = \frac{\overline{a+2d}|^{x+1} - \overline{a+d}|^{x+1}}{d \cdot \overline{x+1}} - \frac{x}{2} d \cdot \overline{a+d}|^{x-1} - \frac{x \cdot \overline{x-1}}{2 \cdot 3} d^2 \cdot \overline{a+d}|^{x-2} \text{ \&c.}$$

$$\overline{a+2d}|^x = \frac{\overline{a+3d}|^{x+1} - \overline{a+2d}|^{x+1}}{d \cdot \overline{x+1}} - \frac{x}{2} d \cdot \overline{a+2d}|^{x-1} - \frac{x \cdot \overline{x-1}}{2 \cdot 3} d^2 \cdot \overline{a+2d}|^{x-2} \text{ \&c.}$$

&c. &c. &c.

Consequently, we get, by addition

$$\underset{\div}{S}^x = \frac{\overline{a+nd}|^{x+1} - a^{x+1}}{d \cdot \overline{x+1}} - \frac{x}{2} d \underset{\div}{S}^{x-1} - \frac{x \cdot \overline{x-1}}{2 \cdot 3} d^2 \underset{\div}{S}^{x-2} \text{ \&c.}$$

$\underset{\div}{S}^x$ being wrote for $a^x + \overline{a+d}|^x + \overline{a+2d}|^x + \overline{a+3d}|^x$ (n),

$\underset{\div}{S}^{x-1}$ for $a^{x-1} + \overline{a+d}|^{x-1} + \overline{a+2d}|^{x-1} + \overline{a+3d}|^{x-1}$ (n),

&c. &c.

From

From whence it follows, that

$$\underset{\div}{S}^{x-1} \text{ is } = \frac{\overline{a+nd}|^{x} - a^{x}}{dx} - \frac{x-1}{2} d\underset{\div}{S}^{x-2} - \frac{x-1 \cdot x-2}{2 \cdot 3} d^{2} \underset{\div}{S}^{x-3} \text{ \&c.}$$

$$\underset{\div}{S}^{x-2} = \frac{\overline{a+nd}|^{x-1} - a^{x-1}}{d \cdot \overline{x-1}} - \frac{x-2}{2} d\underset{\div}{S}^{x-3} - \frac{x-2 \cdot x-3}{2 \cdot 3} d^{2} \underset{\div}{S}^{x-4} \text{ \&c.}$$

&c. &c. &c. &c.

Now these values of $\underset{\div}{S}^{x-1}$, $\underset{\div}{S}^{x-2}$, $\underset{\div}{S}^{x-3}$, &c. being substituted successively, for their respective equals, in the value of $\underset{\div}{S}^{x}$, we find

$$\underset{\div}{S}^{x} = a^{x} + \overline{a+d}|^{x} + \overline{a+2d}|^{x} + \overline{a+3d}|^{x} \, (n) =$$

$$\frac{1}{d \cdot \overline{x+1}} \left\{ \begin{array}{l} \overline{a+nd}|^{x+1} \\ -a^{x+1} \end{array} \right. - \frac{1}{2} \left\{ \begin{array}{l} \overline{a+nd}|^{x} \\ -a^{x} \end{array} \right. + \frac{x\overset{\prime}{A}d}{2} \left\{ \begin{array}{l} \overline{a+nd}|^{x-1} \\ -a^{x-1} \end{array} \right. +$$

$$\frac{x \cdot x-1 \cdot x-2 \cdot \overset{\prime\prime}{A}d^{3}}{2 \cdot 3 \cdot 4} \left\{ \begin{array}{l} \overline{a+nd}|^{x-3} \\ -a^{x-3} \end{array} \right. + \frac{x \cdot x-1 \cdot x-2 \cdot x-3 \cdot x-4 \cdot \overset{\prime\prime\prime}{A}d^{5}}{2 \cdot 3 \cdot 4 \cdot 5 \cdot 6} \left\{ \begin{array}{l} \overline{a+nd}|^{x-5} \\ -a^{x-5} \end{array} \right. \text{ \&c.}$$

$\overset{\prime}{A}$ being $= \frac{1}{2 \cdot 3} = \frac{1}{6}$,

$$\overset{\prime\prime}{A} = \frac{3}{2 \cdot 5} - \frac{4}{2}\overset{\prime}{A} = -\frac{1}{30},$$

$$\overset{\prime\prime\prime}{A} = \frac{5}{2 \cdot 7} - \frac{6}{2}\overset{\prime}{A} - \frac{6 \cdot 5 \cdot 4}{2 \cdot 3 \cdot 4}\overset{\prime\prime}{A} = \frac{1}{42},$$

$$\overset{\prime\prime\prime\prime}{A} = \frac{7}{2 \cdot 9} - \frac{8}{2}\overset{\prime}{A} - \frac{8 \cdot 7 \cdot 6}{2 \cdot 3 \cdot 4}\overset{\prime\prime}{A} - \frac{8 \cdot 7 \cdot 6 \cdot 5 \cdot 4}{2 \cdot 3 \cdot 4 \cdot 5 \cdot 6}\overset{\prime\prime\prime}{A} = -\frac{1}{30},$$

$$\overset{v}{A} = \frac{9}{2 \cdot 11} - \frac{10}{2}\overset{\prime}{A} - \frac{10 \cdot 9 \cdot 8}{2 \cdot 3 \cdot 4}\overset{\prime\prime}{A} - \frac{10 \cdot 9 \cdot 8 \cdot 7 \cdot 6}{2 \cdot 3 \cdot 4 \cdot 5 \cdot 6}\overset{\prime\prime\prime}{A} - \frac{10 \cdot 9 \cdot 8 \cdot 7 \cdot 6 \cdot 5 \cdot 4}{2 \cdot 3 \cdot 4 \cdot 5 \cdot 6 \cdot 7 \cdot 8}\overset{\prime\prime\prime\prime}{A} = \frac{5}{66},$$

&c. &c. &c.

SCHOLIUM. When x is $= -1$, the numerator and denominator of the fraction $\frac{\overline{a+nd}|^{x+1} - a^{x+1}}{d \cdot \overline{x+1}}$ $\left(= \frac{1}{d \cdot \overline{x+1}} \left\{ \begin{array}{l} \overline{a+nd}|^{x+1} \\ -a^{x+1} \end{array} \right. \right)$

both

both vanish.—In that case $\frac{1}{d} \times L : \frac{a+nd}{a}$ is the value of that fraction, as appears by what is already taught.

13.

Writing $2n$ instead of n, we have, by the preceding article,

$$a^x = \overline{a+d}|^x + \overline{a+2d}|^x + \overline{a+3d}|^x (2n) = \frac{1}{d.\overline{x+1}}\begin{cases}\overline{a+2nd}|^{x+1}\\ -a^{x+1}\end{cases}$$

$$-\frac{1}{2}\begin{cases}\overline{a+2nd}|^x\\ -a^x\end{cases} + \frac{x\dot{A}d}{2}\begin{cases}\overline{a+2nd}|^{x-1}\\ -a^{x-1}\end{cases} + \frac{x.x-1.x-2\ddot{A}d^3}{2.3.4}\begin{cases}\overline{a+2nd}|^{x-3}\\ -a^{x-3}\end{cases}$$

&c. and, writing $2d$ instead of d, we have, by the same article,

$$a^x + \overline{a+2d}|^x + \overline{a+4d}|^x + \overline{a+6d}|^x (n) = \frac{1}{2d.\overline{x+1}}\begin{cases}\overline{a+2nd}|^{x+1}\\ -a^{x+1}\end{cases}$$

$$-\frac{1}{2}\begin{cases}\overline{a+2nd}|^x\\ -a^x\end{cases} + \frac{2x\dot{A}d}{2}\begin{cases}\overline{a+2nd}|^{x-1}\\ -a^{x-1}\end{cases} + \frac{2^3.x.x-1.x-2.\ddot{A}d^3}{2.3.4}\begin{cases}\overline{a+2nd}|^{x-3}\\ -a^{x-3}\end{cases}$$

&c.

It appears, therefore, by subtraction, that

$$\overline{a+d}|^x + \overline{a+3d}|^x + \overline{a+5d}|^x (n) \text{ is} = \frac{1}{2d.\overline{x+1}}\begin{cases}\overline{a+2nd}|^{x+1}\\ -a^{x+1}\end{cases}$$

$$-\frac{x\dot{A}d}{2}\begin{cases}\overline{a+2nd}|^{x-1}\\ -a^{x-1}\end{cases} - \frac{\overline{2^3-1}.x.x-1.x-2.\ddot{A}d^3}{2.3.4}\begin{cases}\overline{a+2nd}|^{x-3}\\ -a^{x-3}\end{cases} -$$

$$\frac{\overline{2^5-1}.x.x-1.x-2.x-3.x-4.\dddot{A}d^5}{2.3.4.5.6}\begin{cases}\overline{a+2nd}|^{x-5}\\ -a^{x-5}\end{cases} \text{ \&c.}$$

Schol. When x is $= -1$, the value of $\frac{1}{2d.\overline{x+1}}\begin{cases}\overline{a+2nd}|^{x+1}\\ -a^{x+1}\end{cases}$ is $= \frac{1}{2d} \times L : \frac{a+2nd}{a}$, by what is done above.

14. Taking

14.

Taking the laſt equation from that which immediately precedes it, we have

$$a^x - \overline{a+d}|^x + \overline{a+2d}|^x - \overline{a+3d}|^x + \overline{a+4d}|^x - \overline{a+5d}|^x\,(2n) =$$

$$-\frac{1}{2}\left\{\begin{matrix}\overline{a+2nd}|^x\\ -a^x\end{matrix}\right. + \frac{\overline{2^2-1}\,.\,x\dot{A}d}{2}\left\{\begin{matrix}\overline{a+2nd}|^{x-1}\\ -a^{x-1}\end{matrix}\right. + \frac{\overline{2^4-1}\,.\,x\,.\,x-1\,.\,x-2\,.\,\ddot{A}d^3}{2\,.\,3\,.\,4}$$

$$\left\{\begin{matrix}\overline{a+2nd}|^{x-3}\\ -a^{x-3}\end{matrix}\right. + \frac{\overline{2^6-1}\,.\,x\,.\,x-1\,.\,x-2\,.\,x-3\,.\,x-4\,.\,\dddot{A}d^5}{2\,.\,3\,.\,4\,.\,5\,.\,6}\left\{\begin{matrix}\overline{a+2nd}|^{x-5}\\ -a^{x-5}\end{matrix}\right. \;\&c.$$

n being any poſitive integer.

15.

By reſidual diviſion, ($x - \underset{,}{x}$ being the diviſor, and the values of a and d independent of x,) we get, from Art. 12.

$$a^x \times L : a + \overline{a+d}|^x \times L : \overline{a+d} + \overline{a+2d}|^x \times L : \overline{a+2d} +$$

$$\overline{a+3d}|^x \times L : \overline{a+3d}\;(n)$$

$$= \left\{\begin{matrix}
\dfrac{1}{d\,.\,\overline{x+1}}\left\{\begin{matrix}\overline{a+nd}|^{x+1} \times L : \overline{a+nd}\\ -a^{x+1} \times L : a\end{matrix}\right. - \dfrac{1}{2}\left\{\begin{matrix}\overline{a+nd}|^x \times L : \overline{a+nd}\\ -a^x \times L : a\end{matrix}\right.\\
+ \dfrac{x\dot{A}d}{2}\left\{\begin{matrix}\overline{a+nd}|^{x-1} \times L : \overline{a+nd}\\ -a^{x-1} \times L : a\end{matrix}\right. \;\&c.\\
\dfrac{-1}{d\,.\,\overline{x+1}|^2}\left\{\begin{matrix}\overline{a+nd}|^{x+1}\\ -a^{x+1}\end{matrix}\right. + \dfrac{\dot{A}d}{2}\left\{\begin{matrix}\overline{a+nd}|^{x-1}\\ -a^{x-1}\end{matrix}\right. + \left.\begin{matrix}x-1\,.\,x-2\\ x\,.\,x-2\\ x\,.\,x-1\end{matrix}\right\}\times\\
\dfrac{\ddot{A}d^3}{2\,.\,3\,.\,4}\left\{\begin{matrix}\overline{a+nd}|^{x-3}\\ -a^{x-8}\end{matrix}\right. \;\&c.
\end{matrix}\right.$$

Corollary I. Suppoſing $x = 0$, we find

$$L:a + L:\overline{a+d} + L:\overline{a+2d} + L:\overline{a+3d}\,(n)$$

$$= \begin{cases} \overline{\frac{a}{d} + n - \frac{1}{2}} \times L:\overline{a+nd} - \overline{\frac{a}{d} - \frac{1}{2}} \times L:a, \; -n + \\ \frac{\dot{A}d}{1.2}\left\{\begin{matrix}\overline{a+nd}|^{-1} \\ -a^{-1}\end{matrix}\right. + \frac{\ddot{A}d^3}{3.4}\left\{\begin{matrix}\overline{a+nd}|^{-3} \\ -a^{-3}\end{matrix}\right. + \frac{\dddot{A}d^5}{5.6}\left\{\begin{matrix}\overline{a+nd}|^{-5} \\ -a^{-5}\end{matrix}\right. \text{ \&c.}\end{cases}$$

Corollary II. Taking, in the preceding Corollary, a, d, and n each equal to 1, we have

$$\frac{3}{2} \times L:2 = 1 + \frac{\dot{A}}{2.1.2} + \frac{\overline{2^3-1}.\ddot{A}}{2^3.3.4} + \frac{\overline{2^5-1}.\dddot{A}}{2^5.5.6} \text{ \&c.}$$

16.

As the principal theorem in the laſt article was deduced from Art. 12. ſo from Art. 13. we deduce the following theorem, viz.

$$\overline{a+d}|^x \times L:\overline{a+d} + \overline{a+3d}|^x \times L:\overline{a+3d} + \overline{a+5d}|^x \times L:\overline{a+5d}\,(n)$$

$$= \begin{cases} \frac{1}{2d.\overline{x+1}}\left\{\begin{matrix}\overline{a+2nd}|^{x+1} \times L:\overline{a+2nd} \\ -a^{x+1} \times L:a\end{matrix}\right. - \frac{x\dot{A}d}{2} \times \\ \qquad\qquad \left\{\begin{matrix}\overline{a+2nd}|^{x-1} \times L:\overline{a+2nd} \\ -a^{x-1} \times L:a\end{matrix}\right. \text{ \&c.} \\ \frac{-1}{2d.\overline{x+1}|^2}\left\{\begin{matrix}\overline{a+2nd}|^{x+1} \\ -a^{x+1}\end{matrix}\right. - \frac{\dot{A}d}{2}\left\{\begin{matrix}\overline{a+2nd}|^{x-1} \\ -a^{x-1}\end{matrix}\right. - \left.\begin{matrix}x-1.x-2 \\ x.x-2 \\ x.x-1\end{matrix}\right\} \times \\ \qquad\qquad \frac{\overline{2^3-1}.\ddot{A}d^3}{2.3.4}\left\{\begin{matrix}\overline{a+2nd}|^{x-3} \\ -a^{x-3}\end{matrix}\right. \text{ \&c.}\end{cases}$$

Corollary I. Hence, ſuppoſing $x = 0$, we find

$$L:\overline{a+d} + L:\overline{a+3d} + L:\overline{a+5d} + L:\overline{a+7d}\,(n)$$

=

$$=\begin{cases}\overline{\frac{a}{2d}+n}\times L:\overline{a+2nd}-\frac{a}{2d}\times L:a \\ -n-\frac{\acute{A}d}{1\cdot 2}\left\{\begin{matrix}\overline{a+2nd}|^{-1}\\ -a^{-1}\end{matrix}\right. -\frac{\overline{2^3-1}\cdot\ddot{A}d^3}{3\cdot 4}\left\{\begin{matrix}\overline{a+2nd}|^{-3}\\ -a^{-3}\end{matrix}\right. \\ \qquad -\frac{\overline{2^5-1}\cdot\dddot{A}d^5}{5\cdot 6}\left\{\begin{matrix}\overline{a+2nd}|^{-5}\\ -a^{-5}\end{matrix}\right. \ \&c.\end{cases}$$

COROLLARY II. Taking, in the preceding Corollary, $a=2$, $d=1$, and writing $n-1$ instead of n, we have

$$L:1+L:3+L:5+L:7\,(n)=L:1\times 3\times 5\times 7\,(n)$$

$$=L:2^{n-1}n^n+\begin{cases}1+\frac{\acute{A}}{2\cdot 1\cdot 2}+\frac{\overline{2^3-1}\cdot\ddot{A}}{2^3\cdot 3\cdot 4}+\frac{\overline{2^5-1}\cdot\dddot{A}}{2^5\cdot 5\cdot 6}\ \&c. \\ -n-\frac{\acute{A}}{2\cdot 1\cdot 2\cdot n}-\frac{\overline{2^3-1}\cdot\ddot{A}}{2^3\cdot 3\cdot 4\cdot n^3}-\frac{\overline{2^5-1}\cdot\dddot{A}}{2^5\cdot 5\cdot 6\cdot n^5}\ \&c.\end{cases}$$

$=n\times L:n+\overline{n+\frac{1}{2}}\times L:2,\ -n-\frac{\acute{A}}{2\cdot 1\cdot 2\cdot n}-\frac{\overline{2^3-1}\cdot\ddot{A}}{2^3\cdot 3\cdot 4\cdot n^3}-\frac{\overline{2^5-1}\cdot\dddot{A}}{2^5\cdot 5\cdot 6\cdot n^5}$ &c. $\frac{3}{2}\times L:2$ being $=1+\frac{\acute{A}}{2\cdot 1\cdot 2}+\frac{\overline{2^3-1}\cdot\ddot{A}}{2^3\cdot 3\cdot 4}+\frac{\overline{2^5-1}\cdot\dddot{A}}{2^5\cdot 5\cdot 6}$ &c. by Cor. 2. of the last Article.

Other theorems may, in like manner, be derived from Art. 14. which we may take notice of in an Appendix to this Treatise; and perhaps add some farther improvements on the subject of the last five articles, which some time ago engaged the attention of Mr. DE MOIVRE, Mr. STIRLING, and other eminent mathematicians.

THE

THE

RESIDUAL ANALYSIS.

CHAP. IV.

Of the Properties *of certain* ALGEBRAIC EXPRESSIONS.

THE articles in this chapter will be found of considerable uſe, in geometrical and phyſical enquiries; and, to the end that we may proceed with as much perſpicuity as poſſible, it is thought proper to inſert them here, previous to ſuch enquiries.

I.

Suppoſe E *to be an algebraic expreſſion compoſed of* x *and other quantities; and ſuppoſe, that, how near ſoever* x *be taken to ſome certain quantity* g, E *is poſitive when* x *is leſs than* g, *and negative when* x *is greater than* g; *or poſitive when* x *is greater than* g, *and negative when* x *is leſs than* g: *then ſhall* E, *or its reciprocal, be* = o *when* x *is* = g.

For it is well known, that o is the only limit between poſitive and negative; and it is therefore plain, that either the value of E muſt continually approach to that limit, or increaſe or decreaſe without limit, upon taking *x* nearer and nearer to *g*. If the value of E approaches to o when *x* is taken nearer and nearer

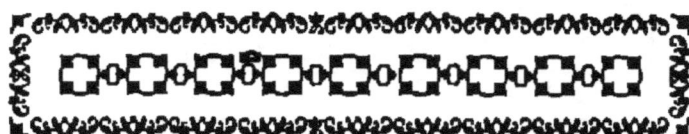

nearer to g, it is evident that E will be $= 0$ when x is $= g$: If the value of E increaſes or decreaſes without limit when x is taken nearer and nearer to g, the value of $\frac{1}{E}$ will then continually approach to o; therefore, $\frac{1}{E}$ being poſitive or negative according as E is poſitive or negative, it is evident that, in ſuch caſe, $\frac{1}{E}$ will be $= 0$ when x is $= g$. Conſequently either E or its reciprocal muſt be $= 0$ when x is $= g$.

2.

Q being an algebraic expreſſion ſo compoſed of x *and other quantities, that neither it nor its reciprocal vaniſhes when* x *is therein taken equal to* g; *the ſaid expreſſion, when* x *is greater or leſs than* g, *between certain limits, will (ſuppoſing it not to become imaginary) be poſitive or negative, according as* (q) *the value of* Q, *when* x *is equal to* g, *is poſitive or negative.*

For, let A and B be poſitive quantities; and ſuppoſe that either Q or $\frac{1}{Q}$ is $= 0$ when x is $= g + A$, and likewiſe that Q or $\frac{1}{Q}$ is $= 0$ when x is $= g - B$, but that neither Q nor $\frac{1}{Q}$ is $= 0$ when x is taken between the limits $g + A$ and $g - B$: then, it is obvious, that, whilſt x is taken between thoſe limits, Q (if it does not become imaginary) will be always poſitive, or always negative; and, conſequently, poſitive or negative according as q is poſitive or negative.

The ſame concluſion follows, if, inſtead of Q or $\frac{1}{Q}$ being $= 0$ when x is $= g + A$, or when x is $= g - B$, thoſe expreſſions then become imaginary, but are real and finite when x is taken between the limits above-mentioned.

3.

Let m be an odd number, or a fraction whoſe numerator and denominator are both odd numbers; and P any algebraic expreſſion ſo compoſed of x and other quantities, that its value ſhall be real when x is either greater or leſs than g, (between certain limits,) and that neither it nor its reciprocal ſhall vaniſh when x is equal to g; alſo let p be the value of P when x is $= g$.

Then, ſeeing that $\overline{x-g}|^m$ is poſitive or negative, according as x is greater or leſs than g; it is evident, (from what is ſaid in the laſt Article) that, how near ſoever x be taken to g, if p be poſitive, $\overline{x-g}|^m \times P$ will be negative when x is leſs than g, and poſitive when x is greater than g; or, if p be negative, the ſame expreſſion $(\overline{x-g}|^m \times P)$ will be negative when x is greater than g, and poſitive when x is leſs than g.

Corollary. Suppoſe E to be an algebraic expreſſion compoſed of x and other quantities: Then, if E be poſitive when x is greater than g, and negative when x is leſs than g, how near ſoever x be taken to g; it is manifeſt, from what we juſt now obſerved, that

$$\overline{x-g}|^m \times Q \text{ may be aſſumed } = E;$$

m being as above ſpecified, and Q ſome algebraic expreſſion conſiſting of ſuch quantities, that its value ſhall be real when x is either greater or leſs than g, (between certain limits,) and that (q) its value when x is equal to g ſhall be finite and poſitive.

Moreover, if E be poſitive when x is leſs than g, and negative when x is greater than g, how near ſoever x be taken to g; it is likewiſe manifeſt, that, in this caſe alſo,

$$\overline{x-g}|^m \times Q \text{ may be aſſumed } = E;$$

m and Q being as before ſpecified, except that q, inſtead of being poſitive, muſt be negative.

Hence it is evident, that, when x is $= g$, (whether q be poſitive or negative,) E or $\frac{1}{E}$ will be $= 0$, according as m is poſitive or negative; which agrees with what is ſaid in Art. 1.

THE

THE

RESIDUAL ANALYSIS.

CHAP. V.

Of the TANGENTS *of* curve Lines.

I.

THE curve AqPq, having its convexity downwards, as in Fig. 1. being referred to the base AB; if the right line NrPr touch the said curve in P; and, brq being parallel to BP, if AB be called x; BP, y; Ab, $x_{,}$; bq, $y_{,}$; and the subtangent BN, s: then will bN be $= s - x + x_{,}$, br $= y - \frac{y}{s} \times \overline{x - x_{,}}$, and the residual b$q$ — br $(= qr) = \frac{y}{s} \times \overline{x - x_{,}} - \overline{y - y_{,}}$. Now, b$rq$ being drawn on either side of BP, bq is manifestly greater than br; therefore $\frac{y}{s} \times \overline{x - x_{,}} - \overline{y - y_{,}}$ (the value of bq — br) must be always positive, when, x being of any given value, $x_{,}$ is either less or greater than x. Fig. 1.

But if the convexity of the curve be upwards, as in Fig. 2. $\frac{y}{s} \times \overline{x - x_{,}} - \overline{y - y_{,}}$ (the value of bq — br) must be always Fig. 2.

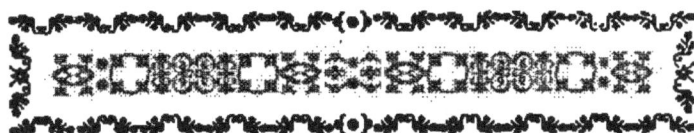

negative, when, x being of any given value, $x_{,}$ is either less or greater than x.

Now; since the expression $\frac{y}{s} \times \overline{x - x_{,}} - \overline{y - y_{,}}$ must be always positive or always negative, when, x being of any given value, $x_{,}$ is either less or greater than x, the quotient of that expression divided by $x - x_{,}$ (viz. $\frac{y}{s} - [x \mid y]$) will, it is obvious, be positive when $x_{,}$ is less than x, and negative when $x_{,}$ is greater than x; or positive when $x_{,}$ is greater than x, and negative when $x_{,}$ is less than x, how near soever $x_{,}$ be taken to x. Therefore, by Chap. 4. Art. 1.

$$\frac{y}{s} - [x \perp y] \text{ will be } = 0\,^{*};$$

$$\text{and, consequently, } s = \frac{y}{[x \perp y]}.$$

Fig. 1. EXAMPLE I. *To draw a tangent to any Parabola, whose equation is* $ax^{\frac{m}{r}} = y$; m *being such, that the convexity of the curve is downwards.*

We have (according to our scheme) $bq - br\ (= qr) = \frac{ax^{\frac{m}{r}}}{s} \times \overline{x - x_{,}} - ax^{\frac{m}{r}} + ax_{,}^{\frac{m}{r}}$; where s must be of such a value, that the value of the whole expression shall be positive, when, x being of any given value, $x_{,}$ is either greater or less than x.

Now if $\frac{ax^{\frac{m}{r}}}{s} \times \overline{x - x_{,}} - ax^{\frac{m}{r}} + ax_{,}^{\frac{m}{r}}$ be always positive when $x_{,}$ is either greater or less than x, $\dfrac{\frac{ax^{\frac{m}{r}}}{s} \times \overline{x - x_{,}} - ax^{\frac{m}{r}} + ax_{,}^{\frac{m}{r}}}{x - x_{,}}$, or its

---

* The other equation (viz. $\dfrac{1}{\frac{y}{s} - [x \perp y]} = 0$) which, if possible, wou'd follow from the same article, is, in this case, manifestly impossible.

Equal

Equal $\frac{a x_{,}^{\frac{m}{r}}}{s} - a x^{\frac{m}{r} - 1} \times \frac{1 + \frac{x_{,}}{x} + \overline{\frac{x_{,}}{x}}\big|^{2} \ (m)}{1 + \overline{\frac{x_{,}}{x}}\big|^{\frac{m}{r}} + \overline{\frac{x_{,}}{x}}\big|^{\frac{2m}{r}} \ (r)}$ (found by our division taught Chap. 2. Art. 1.) it is plain will be positive when $x_{,}$ is less than x, and negative when $x_{,}$ is greater than x. Therefore, by Chap. 4. Art. 1. the value of this last expression will be $= 0$ when $x_{,}$ is therein taken equal to x: Which value is manifestly $= \frac{a x^{\frac{m}{r}}}{s} - \frac{m}{r} . a x^{\frac{m}{r} - 1}$.

Consequently from the equation $\frac{a x^{\frac{m}{r}}}{s} - \frac{m}{r} . a x^{\frac{m}{r} - 1} = 0$, the subtangent s is found $= \frac{r}{m} x$.

In this example we have given the process at full length, that the Reader may the more clearly understand our doctrine.— In future, our examples will, for the most part, be more concise.

EXAMPLE II. *To draw a tangent to the Circle* AqPq, *whose equation is* $2ax - x^2 = y^2$; *the radius* AC *being* $= a$, AB $= x$, BP $= y$. Fig. 3.

From the equation of the curve, we have, by residual division, $a - x = y \times [x \perp y]$; from whence we get $[x \perp y] = \frac{a - x}{y}$. Therefore $\frac{y}{[x \perp y]}$, the value of s or NB, is $= \frac{y^2}{a - x} = \frac{2ax - x^2}{a - x}$.

If BC be called x, and BP as before, the equation of the curve will be $a^2 - x^2 = y^2$; and the subtangent NB will be $= \frac{a^2 - x^2}{x}$.

EXAMPLE III. *To draw a tangent to the Ellipsis* APD, *whose equation is* $b^2 \times \overline{2ax - x^2} = a^2 y^2$; *the semi-transverse axis* AC *being* $= a$, *the semi-conjugate* CD $= b$, AB $= x$, BP $= y$. Fig. 4.

From

From the equation of the curve, we get, by residual division, $ab^2 - b^2x = a^2y[x \perp y]$: hence $[x \perp y]$ is found $= \frac{b^2}{a^2} \times \frac{a-x}{y}$.

Therefore $\frac{y}{[x \perp y]}$, the value of the subtangent NB, is $= \frac{a^2y^2}{b^2 \times \overline{a-x}} = \frac{2ax - x^2}{a-x}$.

If BC be called x, and BP as before; the equation of the curve will be $b^2 \times \overline{a^2 - x^2} = a^2y^2$; and NB will be $= \frac{a^2 - x^2}{x}$.

If, Pd being parallel to AC, Cd be called x; and dP, y: the equation of the curve will be $a^2 \times \overline{b^2 - x^2} = b^2y^2$; and the subtangent $dn = \frac{b^2 - x^2}{x}$.

Fig. 5. EXAMPLE IV. *Let it be proposed to draw a tangent to the Hyperbola* AP, *whose equation is* $b^2 \times \overline{2ax + x^2} = a^2y^2$; *the semi-transverse axis* AC *being* $= a$, *the semi-conjugate* CD $= b$, AB $= x$, BP $= y$.

From the equation of the curve we have $ab^2 + b^2x = a^2y[x \perp y]$: hence we find $[x \perp y] = \frac{b^2}{a^2} \times \frac{a+x}{y}$. Therefore $\frac{y}{[x \perp y]}$, the value of the subtangent NB, is $= \frac{a^2y^2}{b^2 \times \overline{a+x}} = \frac{2ax + x^2}{a+x}$.

If BC be called x, and BP as before; the equation of the curve will be $b^2 \times \overline{x^2 - a^2} = a^2y^2$; and NB will be $= \frac{x^2 - a^2}{x}$.

If, Pd being parallel to AC, Cd be called x; and dP, y; the equation of the curve will be $a^2 \times \overline{b^2 + x^2} = b^2y^2$; and the subtangent $dn = \frac{b^2 + x^2}{x}$.

Fig. 6. Suppose CE, Ce to be the asymptotes of the two branches AP, AQ, of the Hyperbola PAQ: then, BP being parallel to Ce, if CB, BP be called x and y respectively; the equation of the curve will be $xy = p^2$, p being put for $\frac{\sqrt{a^2 + b^2}}{2}$. Therefore,

$y +$

$y + x[x \perp y]$ being $= 0$, $[x \perp y]$ is $= -\frac{y}{x}$, and $\frac{y}{[x \perp y]}$ ($=$ NB) $= -x$; which being negative, indicates, that N is on the contrary side of BP, from C.

EXAMPLE V. *To draw a tangent to the Cissoid* AP, *whose* Fig. 7.
equation is $ay^2 - xy^2 = x^3$.

From the said equation we have $2ay[x \perp y] - y^2 - 2xy[x \perp y] = 3x^2$; and from hence we get $[x \perp y] = \frac{3x^2 + y^2}{2ay - 2xy}$: Which last expression being substituted for $[x \perp y]$ in the equation $s = \frac{y}{[x \perp y]}$, we find $s = \frac{2ay^2 - 2xy^2}{3x^2 + y^2}$.

EXAMPLE VI. *To draw a tangent to the exponential curve, whose equation is* $a^x = y$.

By residual division (see Chap. 3. Art. 9.) we get $a^x \times L:a = [x \perp y]$: Therefore $s \left(= \frac{y}{[x \perp y]}\right)$ is $= \frac{1}{L:a}$, an invariable quantity.

EXAMPLE VII. *To draw a tangent to the exponential curve, whose equation is* $x^x = y$.

From the said equation we find, by residual division, $x^x \times L:x + x^x = [x \perp y]$: Consequently $s \left(= \frac{y}{[x \perp y]}\right)$ is $= \frac{1}{1 + L:x}$.

After the same manner may we draw tangents to any other geometrical, or exponential curve, referred to an axis: but to perform the like by our Analysis, when the curve is a transcendental one, or a spiral, it will be necessary to understand what will be explained in the subsequent part of this chapter.

2.

A line has its concavity turned one way, when the right lines that join any two of its points either fall upon the line itself, or on one side of it, none falling on the opposite side.—Here we comprehend, not only curves, but likewise such lines as have rectilineal parts.

3.

When two lines, having their concavity turned one and the ſame way, have the ſame terms, and one wholly includes the other, the perimeter of that which includes is greater than the perimeter of that which is included *.

4.

Fig. 8. When a curve hqP is convex towards the baſe, and the angle BPN, made by the ordinate BP and tangent PN, is acute; the ordinate bq being drawn, interſecting NP in r, Pr will be greater or leſs than the curve Pq, according as Ab is leſs or greater than AB. For, drawing a tangent to the point q, (between h and P,) interſecting Pr in v; Pvq, by the preceding Article, will be greater than the curve Pq. But rv, which ſubtends an obtuſe angle in the triangle qrv, is greater than qv, which ſubtends an acute angle in the ſame triangle: therefore Pvr (i. e. Pr) will be greater than Pvq; and conſequently Pr ſtill greater than the curve Pq. Moreover, q being on the other ſide of P; Pr, which ſubtends an acute angle in the triangle Pqr, is leſs than the chord Pq, which ſubtends an obtuſe angle in the ſame triangle: therefore, the chord being leſs than the arc it ſubtends, Pr will be ſtill leſs than the curve Pq.

Fig. 9. When the curve hqP is concave towards the baſe, and the angle BPN is acute; the ordinate bq being drawn, interſecting (as before) the tangent NP in r, it is evident, from what is ſaid above, that Pr will be leſs or greater than the curve Pq, according as Ab is leſs or greater than AB.

5.

Fig. 8. 9. Let AB be called x; BP, y; Ab, $\underset{,}{x}$; bq, (parallel to BP,) $\underset{,}{y}$; the ſubtangent BN, s; the tangent PN, t; and the parts hP, hq of the curve, z and $\underset{,}{z}$ reſpectively: then will Pr be $= \frac{t}{s} \times \overline{x - \underset{,}{x}}$ or $\frac{t}{s} \times \overline{\underset{,}{x} - x}$, and the arc P$q = z - \underset{,}{z}$ or $\underset{,}{z} - z$, according as $\underset{,}{x}$

* The laſt two articles are from ARCHIMEDES *de ſphæra et cylindro.*

is

is leſs or greater than x. Therefore, by the laſt Article, if the convexity of the curve be downwards, as in Fig. 8. $\frac{t}{s} \times \overline{x - x_{,}}$ will be greater than $z - z_{,}$, when $x_{,}$ is leſs than x; and $\frac{t}{s} \times \overline{x_{,} - x}$ will be leſs than $z_{,} - z$, when $x_{,}$ is greater than x. Hence it evidently follows, that, the curve being convex towards the baſe, $\frac{t}{s} \times \overline{x - x_{,}}$ will be always greater than $z - z_{,}$, when $x_{,}$ is taken either greater or leſs than x: conſequently the expreſſion $\frac{t}{s} \times \overline{x - x_{,}} - \overline{z - z_{,}}$ will be always poſitive when $x_{,}$ is ſo taken.

Moreover, if the convexity of the curve be upwards, as in Fig. 9. $\frac{t}{s} \times \overline{x - x_{,}}$ being leſs than $z - z_{,}$ when $x_{,}$ is leſs than x, and $\frac{t}{s} \times \overline{x_{,} - x}$ greater than $z_{,} - z$ when $x_{,}$ is greater than x, $\frac{t}{s} \times \overline{x - x_{,}}$ will be always leſs than $z - z_{,}$ when $x_{,}$ is taken either greater or leſs than x: conſequently, in this caſe, the expreſſion $\frac{t}{s} \times \overline{x - x_{,}} - \overline{z - z_{,}}$ will be always negative when $x_{,}$ is ſo taken.

Now, ſince the expreſſion $\frac{t}{s} \times \overline{x - x_{,}} - \overline{z - z_{,}}$ muſt be always poſitive or always negative, when, x being of any given value, $x_{,}$ is either leſs or greater than x; the quotient of that expreſſion divided by $x - x_{,}$ (viz. $\frac{t}{s} - [x \mid z]$) will, it is evident, be poſitive when $x_{,}$ is leſs than x, and negative when $x_{,}$ is greater than x; or poſitive when $x_{,}$ is greater than x, and negative when $x_{,}$ is leſs than x, how near ſoever $x_{,}$ be taken to x. Therefore, by Chap. 4. Art. 1.

$\frac{t}{s}$ —

$$\frac{t}{s} - [x \perp z] \text{ will be } = 0\,^*:$$

and, consequently, $[x \perp z] = \frac{t}{s} = \sqrt{1 + [x \perp y]^2}$;

s being $= \frac{y}{[x \perp y]}$, and $t = \frac{y\sqrt{1+[x \perp y]^2}}{[x \perp y]}$, by what is said in Article 1.

The same conclusion, it is obvious, will likewise hold true, when the ordinates decrease from h towards P.

Corollary I. Hence it is manifest, that the subtangent, tangent, and ordinate, are to each other as 1, $[x \perp z]$, and $[x \perp y]$ respectively.

Corollary II. Suppose hP to be an arc of a circle: then, if x be the versed sine, y the sine, and a the radius thereof; it is well known, the subtangent will be to the tangent, as y to a; and therefore, by the preceding corollary, $y : a :: 1 : [x \perp z]$. Consequently $[x \perp z]$ is then $= \frac{a}{y} = \frac{a}{\sqrt{2ax - x^2}}$.

Moreover, $[x \perp z]$ being $= \frac{[y \perp z]}{[y \perp x]}$ (by Chap. 2. Art. 8.), $\frac{[y \perp z]}{[y \perp x]}$ is $= \frac{a}{y}$, and $[y \perp z] = \frac{a[y \perp x]}{y}$. But, y being $= \sqrt{2ax - x^2}$, we, by residual division, have $1 = \frac{a - x}{\sqrt{2ax - x^2}} \times [y \perp x]$; and, from hence, $[y \perp x] = \frac{\sqrt{2ax - x^2}}{a - x} = \frac{y}{\sqrt{a^2 - y^2}}$. Therefore $[y \perp z]$ is $= \frac{a}{\sqrt{a^2 - y^2}}$.

Let u be the cosine of z; then will $a^2 - y^2$ be $= u^2$; and, consequently, $[u \perp y] = \frac{-u}{y}$. Now $[y \perp z]$ being $= \frac{[u \perp z]}{[u \perp y]}$,

---

* The other equation (viz. $\frac{1}{\frac{t}{s} - [x \perp z]} = 0$) which, if possible, would follow from the same article, is, in this case, impossible.

this

this laſt quantity will be $= \frac{a}{\sqrt{a^2 - y^2}}$; and, therefore, $[u \perp z]$ will be $= \frac{a[u \perp y]}{\sqrt{a^2 - y^2}} = \frac{a[u \perp y]}{u} = -\frac{a}{y} = \frac{-a}{\sqrt{a^2 - u^2}}$.

COROLLARY III. t^2 being $= s^2 + y^2$, we, by reſidual diviſion, have $t[x \perp t] = s[x \perp s] + y[x \perp y]$: Hence $\frac{t}{s} \times [x \perp t] = [x \perp s] + \frac{y}{s} \times [x \perp y]$.

Now AN being called r, s will be $= x - r$; and conſequently $[x \perp s] = 1 - [x \perp r]$: moreover, by what is ſaid above, $\frac{t}{s}$ is $= [x \perp z]$, and $\frac{y}{s} = [x \perp y]$. Therefore $[x \perp z] \times [x \perp t]$ is $= 1 - [x \perp r] + [x \perp y]^2 = [x \perp z]^2 - [x \perp r]$ ($1 + [x \perp y]^2$ being $= [x \perp z]^2$); from whence it appears, that $[x \perp t]$ is $= [x \perp z] - \frac{[x \perp r]}{[x \perp z]}$. Fig. 8.

6.

By means of the concluſions deduced in the preceding article, we are now enabled to apply the theorem we inveſtigated in Art. 1. to tranſcendental curves referred to an axis.

EXAMPLE I. *To draw a tangent to the Cycloid* AP; *whoſe nature is ſuch, that, the ſemicircle* ApD *being deſcribed, and the ordinate* BpP *being drawn perpendicular to the diameter* AD, BP *is* = Bp + (Arc) Ap. Fig. 10.

Let AD be called $2a$; AB, x; BP, y; Bp, u; and the arc Ap, w. Then, y being $= u + w$, $[x \perp y]$ will be $= [x \perp u] + [x \perp w] = [x \perp u] + \frac{a}{u}$. But, by the nature of the circle, u^2 is $= 2ax - x^2$; from whence $[x \perp u]$ is found $= \frac{a - x}{u}$. Therefore, by ſubſtitution, it appears that $[x \perp y]$ is $= \frac{2a - x}{u}$. Conſequently $\frac{y}{[x \perp y]}$, the value of the ſubtangent NB, is $= \frac{uy}{2a - x} = \frac{xy}{u}$.

It is observable, that $\frac{xy}{u}$ (= NB) is to y (= BP), as x (= AB) to u (= Bp): Therefore the tangent NP is parallel to the chord Ap.

Fig. 11. EXAMPLE II. *To draw a tangent to the Quadratrix* APD.

DE being one fourth of the periphery of a circle described about the center C, draw the radius CPG, which suppose equal to a: call AC, b; AB, x; BP, y; EF, v; FG, (parallel to BP,) u; and the arc EG, w. Then, by the nature of the quadratrix, y will be $= \frac{bw}{a}$; from whence we have $[x \perp y] = \frac{b}{a} \times [x \perp w] = \frac{b}{a} \times [v \perp w] \times [x \perp v] = \frac{b[x \perp v]}{u}$, $\frac{[x \perp w]}{[x \perp v]}$ being $= [v \perp w]$ by Chap. 2. Art. 8. and $[v \perp w] = \frac{a}{u}$ by the last article. Moreover, by the nature of the circle, u^2 is $= 2av - v^2$: from whence, by residual division, we get $u[x \perp u] = \overline{a - v} \times [x \perp v]$; and hence $\frac{[x \perp u]}{a - v} = \frac{[x \perp v]}{u}$. It follows therefore, that $[x \perp y]$ is $= \frac{b[x \perp v]}{u} = \frac{b[x \perp u]}{a - v}$; and consequently that $[x \perp v]$ is $= \frac{u[x \perp y]}{b}$, and $[x \perp u] = \frac{a - v}{b} \times [x \perp y]$. Again, by similar triangles, $b - x : y :: a - v : u$; therefore $ay - vy$ is $= bu - ux$. Hence, by residual division, we have $a[x \perp y] - v[x \perp y] - y[x \perp v] = b[x \perp u] - x[x \perp u] - u$; and consequently, by substitution, $a - v - \frac{uy}{b} \times [x \perp y] = \overline{b - x} \times \frac{a - v}{b} \times [x \perp y] - u$. From whence it appears, that $\frac{uy}{b} \times [x \perp y]$ is $= \frac{x \times \overline{a - v}}{b} \times [x \perp y] - u$; therefore $[x \perp y]$ is $= \frac{bu}{uy - x \times \overline{a - v}}$. Consequently $\frac{y}{[x \perp y]}$, the value of the subtangent BN, is $= \frac{y^2}{b} - \frac{uy \times \overline{a - v}}{bu} = \frac{y^2 + x^2 - bx}{b}$.

7. AqPq

Plate I.

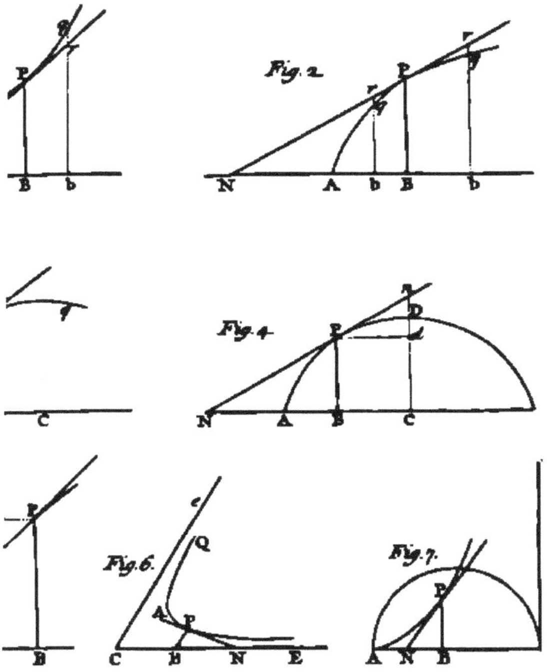

7.

AqPq being a curve of the spiral kind, whose ordinates Cq, CP, Cq, all issue from the point C; let the circular arc defe be described, whose radius is 1; and draw any right line Cd, intersecting the said arc in d. Then, supposing the right line NrPr to touch the curve in P; and supposing Cqr to intersect the said tangent in r, and the circular arc in e: if CP be called y; Cq, $y_{,}$; the sine of the angle dCf, (to the radius 1,) u; the sine of dCe, $u_{,}$; and the sine and cosine of CPN, s and c respectively: the sine of eCf will be denoted by $u\sqrt{1-u_{,}^{2}} - u_{,}\sqrt{1-u^{2}}$ or $u_{,}\sqrt{1-u^{2}} - u\sqrt{1-u_{,}^{2}}$, according as de is less or greater than df; and the sine of CrN, (or CrP,) by $cu\sqrt{1-u_{,}^{2}} - cu_{,}\sqrt{1-u^{2}} - suu_{,} + s\sqrt{1-u^{2}} \times \sqrt{1-u_{,}^{2}}$, which last sine, for brevity sake, we will call k: Moreover Cr will be $= \frac{sy}{k}$, and the residual Cr — Cq ($=qr$) $= \frac{sy}{k} - y_{,} = \frac{sy - ky_{,}}{k}$. Now, Cqr being drawn on either side of CP, Cr is manifestly greater than Cq: therefore $\frac{sy - ky_{,}}{k}$ (the value of Cr — Cq) must be always positive, when, u being of any given value, $u_{,}$ is either less or greater than u. Fig. 12.

If the convexity of the curve be towards C, as in Fig. 13. $\frac{sy - ky_{,}}{k}$ (the value of Cr — Cq) will be always negative, when, u being of any given value, $u_{,}$ is either less or greater than u.

Since the expression $\frac{sy - ky_{,}}{k}$ must be always positive or always negative, when, u being of any given value, $u_{,}$ is either less or greater than u; the quotient of that expression divided by $u - u_{,}$ (viz. $\frac{sy - ky_{,}}{k \cdot \overline{u - u_{,}}}$) will, it is obvious, be positive when $u_{,}$ is less than u,

u, and negative when $u_{,}$ is greater than u; [illegible] $u_{,}$ is greater than u, and negative when $u_{,}$ is less than u, [illegible] ar soever $u_{,}$ be taken to u. Therefore, by Chap. 4. Art. 1. the value of that quotient, when $u_{,}$ is equal to u, (i. e. when $y_{,}$ is $= y$,) or the reciprocal of that value, will be $= 0$. But, by the method taught in Chap. 2. Corollary to Art. 6. the value of $\frac{sy - s_{,}y_{,}}{k \cdot \overline{u - u_{,}}}$, when $u_{,}$ is $= u$, is found equal to $\overline{s[u \perp y] - \frac{cy}{\sqrt{1-u^2}} + s}$.

Therefore $s[u \perp y] - \frac{cy}{\sqrt{1-u^2}}$ is $= 0$*: Hence $\frac{s}{c} = \frac{1}{\sqrt{1-u^2}} \times \frac{y}{[u \perp y]}$.

If the arc df be called w, $\frac{1}{\sqrt{1-u^2}}$ will be $= [u \perp w]$, by Art. 5. It therefore appears, by substitution, that

$$\frac{y[u \perp w]}{[u \perp y]} \text{ is } = \frac{s}{c} = \text{Tang. of the angle CPN.}$$

Suppose CN, perpendicular to the ray CP, to intersect the tangent PN in N: then, radius being to CP, as the tangent of CPN to CN, CN will be $= \frac{y^2[u \perp w]}{[u \perp y]}$.

Seeing c is $= \sqrt{1 - s^2}$, it is evident that $\frac{s}{\sqrt{1-s^2}}$ is $= \frac{y[u \perp w]}{[u \perp y]}$: from whence we have $s = \frac{y[u \perp w]}{\sqrt{[u \perp y]^2 + y^2[u \perp w]^2}}$.

8.

Fig. 12. When the curve AqP is concave towards C, and the angle CPN, made by the ray CP and tangent PN, is acute; the ray Cqr being drawn, intersecting the said tangent in r, Pr will be

---

* The other equation (viz. $\frac{s}{s[u \perp y] - \frac{cy}{\sqrt{1-u^2}}} = 0$) which, if possible, would follow from what is said above, is manifestly impossible.

less

lefs or greater than the curve Pq, according as de is lefs or greater than df: And when the curve AqP is convex towards C, and the angle CPN is acute, as in Fig. 13. Pr will be greater or lefs than the curve Pq, according as de is lefs or greater than df. This may be eafily proved, by reafoning as in Art. 4.

Now, retaining the notation in the laft Art. Pr will be equal to $\frac{y}{k}\times\overline{u\sqrt{1-u_{,}^2}-u_{,}\sqrt{1-u^2}}$ or $-\frac{y}{k}\times\overline{u\sqrt{1-u_{,}^2}-u_{,}\sqrt{1-u^2}}$, according as de is lefs or greater than df. Therefore, calling the parts AP, Aq, of the curve AqPq, z and $z_{,}$ refpectively; it follows that the expreffion $z-z_{,}-\frac{y}{k}\times\overline{u\sqrt{1-u_{,}^2}-u_{,}\sqrt{1-u^2}}$ muft be always pofitive or always negative, when, u being of any value whatever, $u_{,}$ is either greater or lefs than u. It is obvious then, that the quotient of the faid expreffion, divided by $u-u_{,}$ (viz. $\frac{z-z_{,}}{u-u_{,}}-\frac{y}{k}\times\frac{u\sqrt{1-u_{,}^2}-u_{,}\sqrt{1-u^2}}{u-u_{,}}$) will be pofitive when $u_{,}$ is lefs than u, and negative when $u_{,}$ is greater than u; or pofitive when $u_{,}$ is greater than u, and negative when $u_{,}$ is lefs than u, how near foever $u_{,}$ be taken to u. Therefore, by Chap. 4. Art. 1. the value of that quotient, when $u_{,}$ is equal to u, (i.e. when $y_{,}$ is $=y$, and $z_{,}=z$,) or the reciprocal of that value, will be $=0$. But, by the method taught in Chap. 2. Cor. to Art. 6. the value of $\frac{u\sqrt{1-u_{,}^2}-u_{,}\sqrt{1-u^2}}{u-u_{,}}$, when $u_{,}$ is $=u$, is found equal to $\frac{u^2}{\sqrt{1-u^2}}+\sqrt{1-u^2}=\frac{1}{\sqrt{1-u^2}}$; and k is then equal to s. Therefore $[u\perp z]-\frac{y}{s\sqrt{1-u^2}}$ is $=0$*: Hence $[u\perp z]=\frac{y}{s\sqrt{1-u^2}}$.

* The other equation (viz. $\frac{1}{[u\perp z]-\frac{y}{s\sqrt{1-u^2}}}=0$) which, if poffible, would follow from what is faid above, is impoffible.

Seeing

Seeing that $\frac{1}{\sqrt{1-u^2}}$ is $=[u \perp w]$, by Art. 5. it appears, by substitution, that $[u \perp z]$ is $=\frac{y[u \perp w]}{s} = \sqrt{[u \perp y]^2 + y^2[u \perp w]^2}$; s being $=\frac{y[u \perp w]}{\sqrt{[u \perp y]^2 + y^2[u \perp w]^2}}$, by the preceding Article.

From what is said it is evident, that s is $=\frac{y[u \perp w]}{[u \perp z]}$; and consequently c $(=\sqrt{1-s^2})$ is $=\frac{[u \perp y]}{[u \perp z]}$.

Moreover, supposing CQ to be perpendicular to the tangent PQ; radius will be to y, as $\frac{y[u \perp w]}{[u \perp z]}$ $(=s)$ to CQ; therefore CQ will be $=\frac{y^2[u \perp w]}{[u \perp z]}$: and radius will be to y, as $\frac{[u \perp y]}{[u \perp z]}$ $(=c)$ to PQ; therefore PQ will be $=\frac{y[u \perp y]}{[u \perp z]}$.

Corollary. Hence it appears, that the perpendicular CQ, the tangent PQ, and the ray CP, are to each other as $y[u \perp w]$, $[u \perp y]$, and $[u \perp z]$ respectively.

9.

By means of the theorems investigated above, we are now enabled to draw tangents to any spiral whose equation is given.

Fig. 14. Example I. *To draw a tangent to the spiral of* Archimedes; *whose nature is such, that, any circle* AF *being described about the center* C, *and any ray* CfFP *being drawn, the arc* AF *is to* FP *in a constant ratio.*

Let CA be $=m$, C$d=1$, the arc $df=w$, CP $=y$; and let the given ratio of AF to FP be as m to n: Then mw will be $=$AF, and $mw=y-m$. From whence we have $n[u \perp w] = [u \perp y]$; and consequently, by substitution, CN $\left(=\frac{y^2[u \perp w]}{[u \perp y]}\right.$, by Art. 7.) is found $=\frac{y^2}{n}$.

Example

EXAMPLE II. *To draw a tangent to the spiral* CdP; *whose nature is such, that the arc* CdP *is to the ray* CP *in an invariable ratio.* Fig. 15.

Let CdP be $= z$, CP $= y$; and let the invariable ratio of z to y be that of a to b: Then bz will be $= ay$. From whence we get $\frac{a}{b}[u \perp y] = [u \perp z]$, which, by the last Article, is $= \sqrt{[u \perp y]^2 + y^2[u \perp w]^2}$. Hence it appears, that $[u \perp y]$ is $= \frac{b}{\sqrt{a^2 - b^2}} \times y[u \perp w]$; and therefore $\frac{y[u \perp w]}{[u \perp y]}$, the tangent of the angle CPN, is $= \frac{\sqrt{a^2 - b^2}}{b}$, and CN $= \frac{\sqrt{a^2 - b^2}}{b} \times y$.

It is observable, that the angle made by the tangent PN with the ray CP is invariable; which is a known property of the *logarithmic spiral.*

10.

Suppose the moveable curve aB *to revolve along the immoveable curve* AB, *so that the arcs* aB, AB *be always equal; and suppose, that, during the motion, the point* P, *having a certain position with respect to the curve* aB, *describes the curve* OPQ, *the curves and the describing point keeping always in the same plane: then, if from* B, *the point where the two curves,* aB, AB, *touch each other when the describing point is in* P, BP *be drawn;* PR, *perpendicular thereto, shall touch the curve* OPQ *in* P. Fig. 16.

For, about the center B, with the radius BP, describe the circular arc EPF: and, having drawn BFQ, suppose that, when the describing point is in Q, the moveable curve is posited in abD; b being that point thereof which was at B when the describing point was at P, and D being now the point of contact of the two curves abD, ABD: join bB, bQ; and let *ef* touch the curve abD in b, and meet the curve ABD in *f*.—Then, because the arcs BD, bD are equal; and b*f* + *the arc* D*f* is greater than the arc bD, b*f* shall be greater than the arc B*f*, and consequently still greater than the chord B*f*: Wherefore the angle bB*f* (made by bB and the chord B*f*) will be greater than Bb*f*, and Bbe greater than bBg, made by bB and the conti- Fig. 17.

continuation of the chord *f*B. It is evident therefore, that the angle BbQ, which is = B*be* + *eb*Q, will be greater than QBb, which is = bB*e* — QB*e*; and consequently BQ will be greater than bQ. But bQ is manifestly equal to BP, being the same line transferred with the moveable curve. Therefore BQ is greater than BP, i. e. than BE. Hence it appears, that the point Q is without the circular arc EPF.

Fig. 18. Suppose now, that, when the describing point is on the other side of P in O, the moveable curve is posited in *ad*b; b being that point thereof which coincides with B when the describing point is in P, and *d* being now the point of contact of the two curves *ad*b, A*d*B: and join BO, Bb, bO, b*d*.—Then, the arc B*d*, which is equal to the arc b*d*, will be greater than the chord b*d*. But, B*e* being drawn, touching the curve A*d*B in B, and intersecting the chord b*d* in *e*, B*e* + *ed* will be greater than the arc B*d*; and therefore B*e* will be greater than b*e*: Wherefore the angle Bb*e* will be greater than bB*e*. It is plain then, that the angle BbO, which is = Bb*e* + *e*bO, will be greater than bBO, which is = bB*e* — *e*BO; and consequently BO will be greater than bO. But bO is evidently equal to BP, being the same line in a different position. Therefore BO is greater than BP, i. e. than BE. Hence it appears, that the point O is without the circular arc EPF: Therefore, the same being proved of the point Q, it follows, that the said circular arc touches the curve OPQ in P. Consequently PR, which is a tangent to that circular arc, will also touch the curve OPQ in P.

Whatever position the describing point P may have with respect to the moveable curve, and whether that curve revolves along the convexity or concavity of the immoveable one, the tangents to the curve described by P will always be determined by the rule here given; and in no case will the demonstration differ greatly from the above.—The describing point may indeed be so posited with respect to the moveable curve, that the circular arc EPF shall sometimes fall without the curve OPQ, which occasions some little difference in the demonstration; but there will not then be any particular difficulty in it: Any farther explanation is therefore unnecessary.

This article (which was suggested by one in Huygen's Horolog. Oscillator.) *will be found of great use in many enquiries concerning* cycloidal curves.

Scho-

SCHOLIUM. If the point P, instead of keeping a certain position with respect to the curve aB, be supposed to move in such a manner along the base MN of that curve, whilst the curve itself revolves as before-mentioned, that any ordinate being drawn from the point of contact B perpendicular to the said base, P shall always be at the end of the correspondent abscissa; then, by what is proved above, will that abscissa (or the base) be a tangent to the curve OPQ described by the point P during such motion. Fig. 19.

11.

With respect to curves referred to an axis, it is obvious, that when the convexity of the curve is downwards, as in Fig. 1. the value of the quotient of the ordinate divided by the subtangent increases as the abscissa (x) is taken greater and greater: and when the convexity is upwards, as in Fig. 2. the value of the said quotient decreases as x is taken greater and greater. Now, by what is said above, that quotient (resuming the notation in Art. 1.) is equal to $[x \perp y]$. Therefore, in the former case, $[x \perp y] - [x_{,} \perp y_{,}]$ will be positive when $x_{,}$ is less than x, and negative when $x_{,}$ is greater than x: and, in the latter case, $[x \perp y] - [x_{,} \perp y_{,}]$ will be positive when $x_{,}$ is greater than x, and negative when $x_{,}$ is less than x. It is evident then, that, $x_{,}$ being either less or greater than x, $\dfrac{[x \perp y] - [x_{,} \perp y_{,}]}{x - x_{,}}$ will be always positive or always negative, according as the convexity is downwards or upwards. Hence it is plain, that the value of $[x \perp\!\!\perp y]$ (supposing both it and its reciprocal to be finite) will be positive or negative, according as the convexity of the curve is downwards or upwards. Fig. 1. 2.

12.

With regard to curves referred to a fixed point C, it is evident, that, when the concavity of the curve is towards C, as in Fig. 12. the perpendicular from C on the tangent increases as the ray or ordinate (y) is taken greater and greater: and when the convexity is towards C, as in Fig. 13. the said perpendicular decreases Fig. 12. 13.

as y is taken greater and greater. But, by what is said above, that perpendicular (resuming the notation in Art. 7. and 8.) is equal to $\frac{y^2[u \perp w]}{[u \perp z]}$. Consequently, in the former case, $\frac{y^2[u \perp w]}{[u \perp z]} - \frac{y_{,}^2[u_{,} \perp w_{,}]}{[u_{,} \perp z_{,}]}$ will be positive when $u_{,}$ is less than u, and negative when $u_{,}$ is greater than u: and, in the latter case, $\frac{y^2[u \perp w]}{[u \perp z]} - \frac{y_{,}^2[u_{,} \perp w_{,}]}{[u_{,} \perp z_{,}]}$ will be positive when $u_{,}$ is greater than u, and negative when $u_{,}$ is less than u. Therefore it is manifest, that, $u_{,}$ being either less or greater than u, $\overline{\frac{y^2[u \perp w]}{[u \perp z]} - \frac{y_{,}^2[u_{,} \perp w_{,}]}{[u_{,} \perp z_{,}]}} \div \overline{u - u_{,}}$ will be always positive or always negative, according as the concavity or convexity is towards C. From whence it appears, that the value of $\left[u \perp \frac{y^2[u \perp w]}{[u \perp z]}\right]$ (supposing both it and its reciprocal to be finite) will be positive or negative, according as the curve is concave or convex towards C.

Plate II.

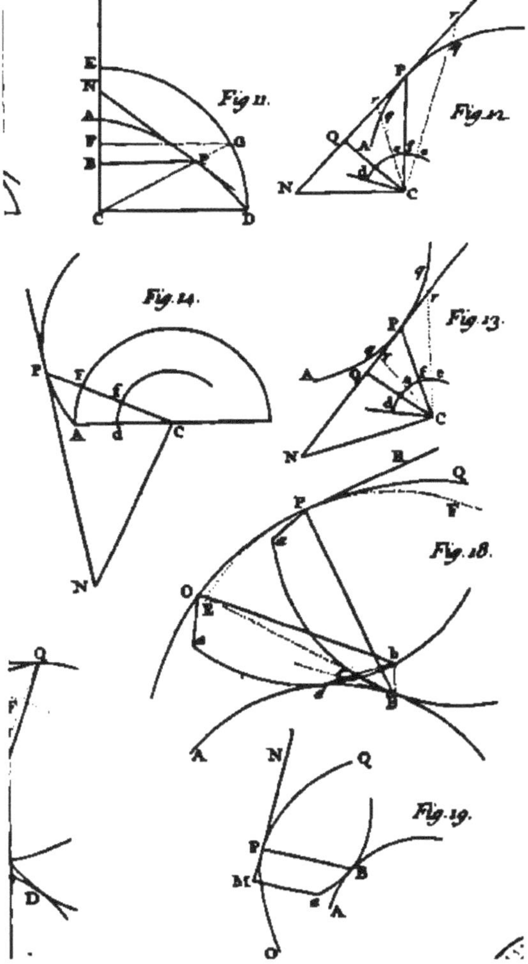

Fig. 11.
Fig. 12.
Fig. 14.
Fig. 13.
Fig. 18.
Fig. 19.

THE

RESIDUAL ANALYSIS.

CHAP. VI.

Of the Investigation of useful Theorems, *by finding the nature of a Curve from a given property of its* Tangents.

I.

BF *Being parallel to* AE; *it is proposed to find a curve line, such, that,* ab *being a tangent thereto, at any point thereof,* $\overline{Bb}|^2 - \overline{Aa}|^2$ $(= \overline{Bb + Aa} \times \overline{Bb - Aa})$ *shall always be equal to the invariable square* c^2: *which tangent is supposed to intersect* AE *and* BF *in the points* a *and* b *respectively.* Fig. 20.

Supposing p to be the point where ab touches the required curve, let mp be drawn parallel to AE; and call AB, a; Am, x; mp, y.

Then will Aa be $= y - x\,[x \perp y]$,

$$Bb = y + \overline{a - x} \times [x \perp y],$$

and $\overline{Bb}|^2 - \overline{Aa}|^2 = \overline{a^2 - 2ax} \times [x \perp y]^2 + 2ay\,[x \perp y] = c^2$.

 Hence,

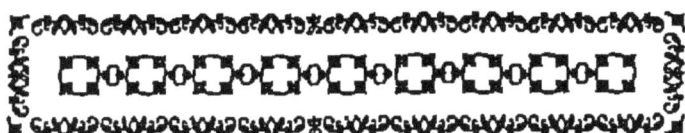

Hence, by residual division, we get

$$\overline{a^2 - 2ax} \times [x \perp y]^2 \times [x \perp y] + ay[x \perp y] = 0;$$

from whence we have $[x \perp y] = \frac{y}{2x - a}$.

Which last quantity being substituted for its equal in the value of c^2, we find

$$y^2 = \frac{c^2}{a} \times \overline{2x - a}.$$

The curve therefore is a Parabola; AB coincides with a diameter thereof, the parameter of which is $\mp \frac{2c^2}{a}$; and, V being the point where that diameter meets the curve, AV is $=$ BV.

Corollary. AB being a diameter of any conical Parabola, and aAa′ parallel to the corresponding ordinates; if Aa on one side of A be equal to Aa′ taken on the contrary side of A, the tangents ap, a′p′ will always intersect each other in the right line Bb, being that ordinate continued whose abscissa VB is $=$ VA.

2.

Fig. 21. BF *being parallel to* AE, *it is proposed to find a curve line, such, that,* ab *being a tangent thereto, at any point thereof,* $\overline{Bb}^2 + \overline{Aa}^2$ *shall always be equal to the invariable square* c^2.

Let the lines AB, A*m*, *mp* be as in the preceding article: Then Aa being $= y - x[x \perp y]$, and Bb $= y + \overline{a - x} \times [x \perp y]$; $\overline{Bb}^2 + \overline{Aa}^2$ is $= 2 \times \overline{y - x[x \perp y]}^2 + \overline{a^2 - 2ax} \times [x \perp y]^2 + 2ay[x \perp y] = c^2$.

Hence, by residual division, and dividing by $2[x \perp y]$, we get

$$2x^2[x \perp y] - 2xy + \overline{a^2 - 2ax} \times [x \perp y] + ay = 0;$$

from whence we have $[x \perp y] = \frac{ay - 2xy}{2ax - a^2 - 2x^2}$.

Which

Which last quantity being substituted for its equal in the value of c^2, it appears that the equation of the curve is

$$y^2 = \frac{c^2}{a^2} \times \overline{a^2 - 2ax + 2x^2}:$$

Answering to an Hyperbola; whereof CV equal to $\frac{c}{\sqrt{2}}$, and parallel to AE and BF, is a semi-diameter; whose semi-conjugate is $AC = BC = \frac{a}{2}$; C being the center.

As by investigating the first proposition, we discovered a remarkable property of the Parabola; so here we discover a property of the Hyperbola, equally remarkable: And it is obvious, that a Corollary may be drawn from what is done in this article, similar to that in the preceding.

3.

BF *being still parallel to* AE; *it is proposed to find a curve line, such, that,* ab *being a tangent thereto, at any point thereof, the rectangle* Aa × Bb *shall always be equal to the invariable square* c^2. Fig. 21. 23.

The lines AB, Am, mp being as in the preceding articles; Aa will be $= y - x[x \perp y]$, and $Bb = a[x \perp y] \pm \overline{y - x[x \perp y]}$.

Therefore Aa × Bb will be $= a[x \perp y] \times \overline{y - x[x \perp y]} \pm \overline{y - x[x \perp y]}|^2 = c^2$.

Hence, by means of our residual division, we get

$$\overline{a \mp 2x} \times \overline{y - x[x \perp y]} - ax[x \perp y] = 0;$$

from whence we have $[x \perp y] = \frac{ay \mp 2xy}{2ax \mp 2x^2}$.

Consequently, by substitution, we find

$$Aa = \frac{ay}{2a \mp 2x},\ Bb = \frac{ay}{2x},\ \text{and}\ \frac{a^2y^2}{4 \cdot \overline{ax \mp x^2}} = c^2.$$

Therefore, in the first case, the equation being $y^2 = \frac{4c^2}{a^2} \times \overline{ax - x^2}$, the curve is a Circle or an Ellipsis; and in the second case, the equation being $y^2 = \frac{4c^2}{a^2} \times \overline{ax + x^2}$, the curve is an Hyperbola. AE, BF touch the conic section, or opposite hyperbolas, in A and

and B: AB being a diameter of the section; whose conjugate is equal to $2c$, and parallel to those tangents.

Corollary I. It appears that c, the semi-conjugate to the diameter AB, is a mean proportional between Aa and Bb.

Corollary II. x being to $\frac{y}{2}$ as a to $\frac{ay}{2x}$, the value of Bb; it follows, that if AE, BF be parallel tangents to any ellipsis, or opposite hyperbolas, and mp any ordinate to the diameter AB, a line drawn from the point of contact A, so as to bisect mp, will always meet the tangent from p in the line BF.

Corollary III. Let $a'b'$ be another tangent to the curve. Then, Aa' being to Bb as Aa to Bb', the right lines $b'a$, ba' will meet in the diameter AB, or in the continuation thereof.

Scholium. If it be required to describe, to a given center, a conic section, or opposite hyperbolas, that shall touch three right lines given by position: Draw a line parallel to one of the given tangents, so that the given center shall be between them and equidistant from each, which line it is plain must be another tangent to the required curve; and then the diameter corresponding to the two points of contact of the parallel tangents may be readily found by this Corollary, and its conjugate will be known by Cor. I. The description of the curve then easily follows.

Corollary IV. By considering BF to be removed to an infinite distance from AE, we may infer from the last Corollary, that, if AE, ab, $a'b'$ be tangents to any conical parabola, right lines drawn through a and a' parallel to $a'b'$, ab respectively, will meet in the continuation of that diameter AB which passes through the point where AE touches the curve.

Scholium. This Corollary, it is obvious, enables us to describe, with great facility, a parabola that shall touch three right lines given by position, and have its axis parallel to another right line given by position; and likewise to describe a parabola, that shall touch four right lines given by position.

4. AB,

4.

AB, BC *making any angle at* B; *it is proposed to find a curve line, such, that,* ab *being a tangent thereto, at any point thereof, the rectangle* Aa × bC *shall always be equal to the invariable square* c^2. Fig. 24. 25.

Supposing p to be the point where ab touches the required curve, let AD, mp be drawn parallel to BC, AB respectively; and call AB, a; BC, b; Am, x; mp, y.

Then will Aa be $= y - x[x \perp y]$, Ba $= a - \overline{y - x[x \perp y]}$,

$$\text{Bb} = \frac{a - \overline{y - x[x \perp y]}}{[x \perp y]}, \text{ and bC} = b \backsim \frac{a - \overline{y - x[x \perp y]}}{[x \perp y]}.$$

Therefore

$$\text{Aa} \times \text{bC is} = b \times \overline{y - x[x \perp y]} \backsim \frac{a \times \overline{y - x[x \perp y]} - \overline{y - x[x \perp y]}^2}{[x \perp y]} = c^2.$$

Hence, after multiplying by $[x \perp y]$, we, by means of our residual division, readily find

$$[x \perp y] = \frac{ax + by - 2xy \mp c^2}{2bx - 2x^2}.$$

Consequently, by a proper substitution, it appears that the equation of the curve is

$$\overline{ax + by \mp c^2}^2 = \overline{4ab \mp 4c^2} \times xy:$$

Which corresponds to an Ellipsis, or Hyperbola. And if CD be parallel to BA; AB, BC, CD, and DA will be tangents to the ellipsis, or opposite hyperbolas. Moreover, taking AE and CF each equal to $\frac{c^2}{b}$, the right line EF will be a diameter of the Figure, whose conjugate diameter will be $= \frac{2c}{b} \times \sqrt{ab \mp c^2}$.

COROLLARY. The middle point P of the right line AC is the center of the Figure: And, (supposing $a'b'$ another tangent to the curve,) Aa being to Cb', as Aa' to Cb; if ab', a'b be joined, and those lines bisected, a right line drawn through the points of bisection will pass through the point P.

For

Fig. 26. 27. For let Bd be to BC, as Aa to Cb′; and parallel to Cd draw b′e, bf; alſo, having biſected thoſe three parallels in m, n, and o, and the three lines AC, ab′, a′b in p, q, r reſpectively, draw mp, nq, or, which will all three be parallel to AB. Then from the analogies

$$\text{de} : \text{Cb}' :: \text{Bd} : \text{BC} :: \text{Aa} : \text{Cb}',$$

$$\text{and df} : \text{Cb} :: \text{Bd} : \text{BC} :: \text{Aa}' : \text{Cb},$$

it appears that de will be = Aa, and df = Aa′.

Therefore, in caſe the firſt (Fig. 26.) ae and a′f will each be equal to Ad; and the parallels mp, nq, or each equal to ½Ad. Conſequently, mno being a right line, pqr is, in this caſe, a right line parallel thereto.

Moreover, in caſe the ſecond (Fig. 27.) ae and a′f will be equal to Ad + 2de and Ad + 2df reſpectively; and the parallels mp, nq, or equal to ½Ad, ½Ad + de, and ½Ad + df reſpectively. It is evident therefore, that, in this caſe, pqr is a right line parallel to dC.

Scholium. If it be required to deſcribe an ellipſis, or oppoſite hyperbolas, that ſhall touch five right lines given by poſition; the center of the Figure may be eaſily found by this Corollary: And then we may proceed according to the Scholium to Cor. 3. of the laſt article.

5.

Fig. 28. AB, BC *making any angle at* B; *it is propoſed to find ſuch a curve line, that,* ab *being a tangent thereto, at any point thereof,* Aa × bC *ſhall always be to* Ba × Bb *in the invariable ratio of* a^2 *to* c^2.

Lines being drawn and denominated as in the preceding article; we have, from what is there ſaid, and from the given ratio of the rectangle Aa × bC to the rectangle Ba × Bb,

$$c^2 \times \overline{\overline{b[x \perp y] - a} \times \overline{y - x[x \perp y]} + \overline{y - x[x \perp y]}}\,^2 = a^2 \times \overline{a - y + x[x \perp y]}\,^2.$$

Hence,

Hence, by means of our residual division, ($x - \underset{\cdot}{x}$ being the divisor,) we get

$$[x \perp y] = \frac{bc^2y - a\,.\,\overline{2a^2 - c^2}\,.\,x + \overline{2a^2 - 2c^2}\,.\,xy}{2bc^2x + \overline{2a^2 - 2c^2}\,.\,x^2}.$$

Which last expression being substituted for its equal in the equation above, the equation of the curve will be obtained. Or the same may be found (perhaps more readily) by proceeding as follows.

Let $\frac{1}{[y \perp x]}$ be substituted for its equal $[x \perp y]$, in the first equation, and there will result

$$c^2 \times \overline{\overline{b - a\,[y \perp x]} \times \overline{y\,[y \perp x] - x} + \overline{y\,[y \perp x] - x}\,}^2 =$$
$$a^2 \times \overline{\overline{a - y}\,.\,[y \perp x] + x}^2.$$

From whence, by residual division, ($y - \underset{\cdot}{y}$ being the divisor,) we find

$$[y \perp x] = \frac{bc^2y - a\,.\,\overline{2a^2 - c^2}\,.\,x + \overline{2a^2 - 2c^2}\,.\,xy}{2a^4 - 2a\,.\,\overline{2a^2 - c^2}\,.\,y + \overline{2a^2 - 2c^2}\,.\,y^2}.$$

Therefore, $[x \perp y]$ being $= \frac{1}{[y \perp x]}$,

$$\frac{bc^2y - a\,.\,\overline{2a^2 - c^2}\,.\,x + \overline{2a^2 - 2c^2}\,.\,xy}{2bc^2x + \overline{2a^2 - 2c^2}\,.\,x^2} \text{ will be } = \frac{2a^4 - 2a\,.\,\overline{2a^2 - c^2}\,.\,y + \overline{2a^2 - 2c^2}\,.\,y^2}{bc^2y - a\,.\,\overline{2a^2 - c^2}\,.\,x + \overline{2a^2 - 2c^2}\,.\,xy}.$$

And the equation of the curve, from thence found, is

$$a^2c^2x^2 + b^2c^2y^2 + 2ab\,.\,\overline{2a^2 - c^2}\,.\,xy - 4a^2bx = 0;$$

which corresponds to an Ellipsis or Hyperbola, according as c is greater or less than a.—It appears that AB, and BC will touch the Figure in A and C: And that if CE be parallel to BA, and equal to $\frac{2a^2 - c^2}{c^2}\,.\,a$; AF, parallel to BE, will coincide with a diameter of the conic section; which diameter will be equal to $\frac{c^2}{c^2 \backsim a^2}$BE, and its conjugate equal to $\frac{2a^2}{\sqrt{c^2 \backsim a^2}}$.

COROLLARY I. This conclusion suggests some remarkable properties of the conic sections; and also an easy method of de-

 scribing

ſcribing elliptical or hyperbolic trajectories, that ſhall touch right lines given by poſition, in certain caſes, to which the theorems in the preceding articles are not ſo readily, if at all applicable.

EXAMPLE. *To deſcribe a trajectory that ſhall touch three right lines given by poſition, and two of them in given points.*

Fig. 29. Let AB, BC, ab be the three given lines, and A and C the points of contact of the two firſt of thoſe lines.

On the line BA take Ae equal to AB, and parallel thereto draw C*f*; alſo, having drawn eb*f*, draw *f*Bg meeting Cag in g: then will gA continued coincide with a diameter of the required trajectory. In the ſame manner may the direction of the diameter from C be found; and conſequently the center of the conic ſection, it being the point where thoſe diameters interſect each other.—The buſineſs then may be eaſily completed, after the manner commonly taught by the writers on Conics.

This conſtruction is ſo eaſily inferred from what is done above, that I think it unneceſſary to be more explicit.

Fig. 28. COROLLARY II. If *c* be equal to *a*, the required curve will be a Parabola. It therefore evidently follows, that, any two tangents AB, BC being drawn to any conical parabola A*p*C, touching the ſame at A and C, and interſecting each other at B; if any third tangent ab be drawn (to the ſame parabola) interſecting the tangents AB, BC, at a and b reſpectively, $Aa \times Cb$ will be $= Ba \times Bb$. And a knowledge of this remarkable property of the parabola enables us to find ſome others, and readily to perform the buſineſs mentioned in the Scholium to Cor. 4. Art. 3.

The ſolutions to theſe problems being eaſily obtained by means of our reſidual diviſion, and the firſt theorem in the preceding chapter, without any farther knowledge of our doctrine; they are, it is preſumed, not improperly inſerted here.—In a ſubſequent chapter, we ſhall ſhew how, by a different artifice in our Calculus, ſome other theorems relating to curve lines may be inveſtigated from given properties of their tangents.

Some of the theorems here inveſtigated may be ſeen, demonſtrated in a different manner, in *Sir* ISAAC NEWTON'*s Philoſ. Natur. Princ. Mathem.*

THE

Plate III.

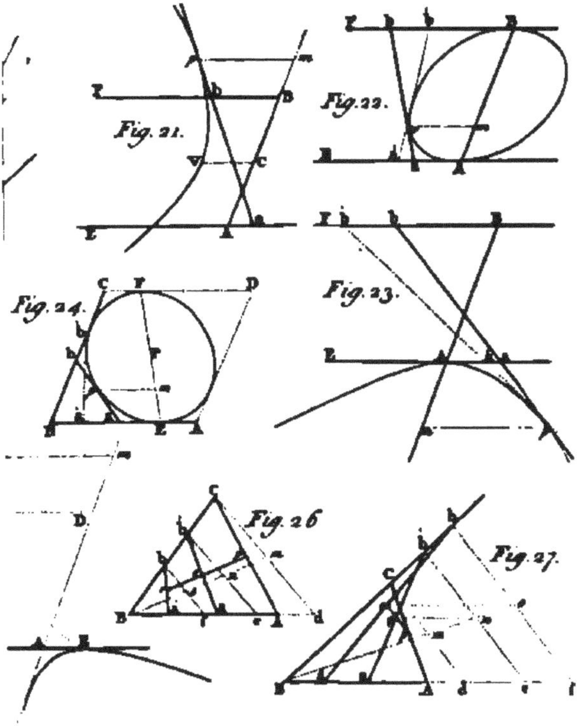

THE

RESIDUAL ANALYSIS.

CHAP. VII.

Of the Evolution *and* Curvature *of Lines*; *with some Inferences relating to the* Focuses *of reflected and refracted rays, and the curves call'd* Caustics.

I.

A Perfectly flexible thread, *dCaA*, being applied along the convexity of the curve *dCa*, from *d* to *a*; suppose the part *aA* (of such thread) to be extended in a right line that touches the said curve in *a*; suppose also, that, whilst one end of the thread remains fixed at *d*, the other end A be moved towards D, (in the same plane with the curve,) so that the thread be continually unwrapped from the curve, and the part CP which is disengaged therefrom be always extended in a right line that touches the curve: then shall the point A trace the *involute* curve APD that is said to be described by the evolution of *aCd*, which is itself called the *evolute*, and the right line CP is called the *radius of evolution* corresponding to the point P. Fig. 30.

2.

Let P*r* be drawn at right angles to CP; and, with the radius CP, describe the circle EPF. Draw any other radius of evolu-

 tion

tion *eq*, *e* being between *a* and C; join C*r*; and draw C*qr*, intersecting P*r* in *r*. Then, by the nature of the evolution, *eq* + *the curve* C*e* being manifestly = CP, *eq* + *the chord* C*e* is less than CP; consequently the right line C*q* is still less than CP. But C*r* is greater than CP: therefore C*q* is less than C*r*, and the involute AP is wholly between the right line P*r* and the evolute *a*C.

The point *e* being on the other side of C, and *g* being the point where PC continued intersects *er*; *eq* (= CP + *the curve* C*e*) is less than *eg* + *g*P: therefore *gq* is less than *g*P. But *gr* is greater than *g*P: therefore *gq* is less than *gr*, and the involute P*q* is wholly between the right line P*r* and the evolute C*e*. Consequently both P*r*, and the circle EPF, touch the involute in P.

The point *e* being between C and *d*, suppose F to be the point where the circle EPF intersects *eqr*. Then *eq* (= CP + *the curve* C*e* = CF + *the curve* C*e*) will be greater than *e*F: consequently the circular arc PF falls within the involute P*q*; and it is plain, that no circle described through P, with a radius less than CP, can pass between PF and P*q*. Moreover, *gq* being less than *g*P, (as before observed,) a circle described from any point *g* through P, with a radius *g*P greater than CP, passes without P*q* and PF. Therefore no circle described through P can pass between P*q* and PF.

The point *e* being between C and *a*, CP (as is proved above) is greater than C*q*: consequently the circular arc PE falls without the involute P*q*; and it is evident, that no circle described through P, with a radius greater than CP, can pass between PE and P*q*. Moreover, *f* being the point where *qe* continued intersects CP, P*f* + *f*C (= *qe* + *the curve e*C) being less than *qf* + *f*C, P*f* is less than *qf*: consequently a circle described from any point *f* through P, with a radius *f*P less than CP, passes within P*q* and PE. It follows therefore, that no circle described through P can pass between the circle EPF and the involute curve APD, all other circles passing either within or without both the said circle and curve.

3.

The circle which touches a curve so closely, that no circle can be drawn through the point of contact between them, is said to have the same curvature with the curve at that point. Which circle

circle is called the *circle of curvature*; its center, the *center of curvature*; and its semidiameter, the *radius of curvature*, corresponding to such point of contact.

It appears then, that the circle EPF (whose center is C) is the circle of curvature of the involute APD, at P. Therefore, by considering any curve as an involute, the radius of curvature (or evolution), at any point thereof, may be readily found, as in the following articles.

4.

The curve APD being referred to a base AK, suppose *aCd* to be the curve by whose evolution APD is described: draw CH Fig. 31.
parallel to KA; and suppose GH, parallel to the ordinate PB, to intersect the said base in I. Call AB, x; BP, y; the curve AP, z; the tangent NP, t; the subtangent BN, s; AI, b; GI, c; GH, v; CH, u; the curve *a*C, w; and (CP) the radius of curvature at P, R.

Then, by similar triangles and Chap. 5. Art. 5.

$$s : t :: v + y + c : \mathrm{R} :: 1 : [v \perp \mathrm{R}] \; (= [v \perp w]).$$

Therefore 1 is $= \frac{s}{t} \times [v \perp \mathrm{R}]$, and $\frac{\mathrm{R}s}{t} - y - c = v$.

From the last equation, we get, by residual division,

$$\frac{s[v \perp \mathrm{R}]}{t} + \frac{\mathrm{R}[v \perp s]}{t} - \frac{\mathrm{R}s\,[v \perp t]}{t^2} - [v \perp y] = 1 = \frac{s[v \perp \mathrm{R}]}{t}.$$

Hence R is found $= \frac{t^2[v \perp y]}{t[v \perp s] - s[v \perp t]} = \frac{t^2[x \perp y]}{t[x \perp s] - s[x \perp t]}$ by Chapter 2. Article 8.

But $\frac{t[x \perp s] - s[x \perp t]}{t^2}$ is $= \left[x \perp \frac{s}{t}\right] = \left[x \perp \frac{1}{[x \perp z]}\right] = -\frac{[x \perp z]}{[x \perp z]^2} = -\frac{[x \perp y] \times [x \perp y]}{[x \perp z]^3}$. Therefore, by substitution, we have

$$\mathrm{R} = -\frac{[x \perp y] \times [x \perp z]^3}{[x \perp z]} = -\frac{[x \perp z]^3}{[x \perp y]},$$

where $[x \perp z]$ is $= \sqrt{1 + [x \perp y]^2}$.

From

From those equations, and what is proved in Chap. 2. Art. 9. other expressions for the value of R may be obtained, viz.

$$\mathrm{R} = \frac{[y \dot{\perp} x] \times [y \dot{\perp} z]^2}{[y \ddot{\perp} z]} = \frac{[y \dot{\perp} z]^3}{[y \ddot{\perp} x]};$$

$$\text{and } \mathrm{R} = \frac{[z \dot{\perp} y]}{[z \ddot{\perp} x]} = -\frac{[z \dot{\perp} x]}{[z \ddot{\perp} y]}.$$

Or the same may be obtained from these analogies, viz.

$$y : t :: u + b - x : \mathrm{R} :: [v \dot{\perp} u] : [v \dot{\perp} \mathrm{R}] \, (= [v \dot{\perp} w]),$$

which follow from similar triangles and Chap. 5. Art. 5.

Example. *In the Parabola,* ax being $= y^2$, x is $= \frac{y^2}{a}$, $[y \dot{\perp} x] = \frac{2y}{a}$, $[y \ddot{\perp} x] = \frac{2}{a}$, and $[y \dot{\perp} z]$ $(= \sqrt{1 + [y \dot{\perp} x]^2})$ $= \sqrt{1 + \frac{4y^2}{a^2}}$. Therefore $\mathrm{R} = \frac{[y \dot{\perp} z]^3}{[y \ddot{\perp} x]}$ is, in such curve, equal to $\frac{\overline{a^2 + 4y^2}|^{\frac{3}{2}}}{2a^2}$.

In the Ellipsis and Hyperbola $2amx \mp ax^2$ is $= 2my^2$: and by residual division, and substitution,

$$\mathrm{R} \text{ is found } = \frac{\overline{a^2m^2 + 4m^2y^2 \mp 2amy^2}|^{\frac{3}{2}}}{2a^2m^2}:$$

a being the Parameter, and m the semi-transverse Axis.

It is observable, that, in each of the Conic Sections, the Radius of Curvature, at the Vertex of the Figure, is equal to *Half the Parameter.*

5.

COROLLARY.

The equation of the involute curve being given, we may, from thence and what is done above, readily find the nature of the evolute. For, from what is said in the last article, we have

$$v = \frac{s\mathrm{R}}{t} - y - c = \frac{\mathrm{R}}{[x \dot{\perp} z]} - y - c \begin{cases} = -\dfrac{[x \dot{\perp} z]^2}{[x \ddot{\perp} y]} - y - c, \\ = \dfrac{[y \dot{\perp} x] \times [y \dot{\perp} z]^2}{[y \ddot{\perp} x]} - y - c; \end{cases}$$

$u =$

$$u=\frac{yR}{t}+x-b=\frac{R}{[y\perp x]}+x-b\begin{cases}=\frac{[y\perp x]^2}{[y\dot{\perp} x]}+x-b,\\ =-\frac{[x\perp y]\times[x\perp x]^2}{[x\dot{\perp} y]}+x-b.\end{cases}$$

From whence, by means of the given equation, x and y may be expunged: Confequently the relation between v and u will then appear.

EXAMPLE. *Let the involute be the conical Parabola, whofe equation is* $ax=y^2$.

Then $[y\perp x]$ being $=\frac{2y}{a}$, $[y\dot{\perp} x]=\frac{2}{a}$, and $[y\perp x]^2=1+\frac{4y^2}{a^2}$;

$$v \text{ is } =\frac{4y^3}{a^2}-c,\ \text{and } u=\frac{a}{2}+\frac{3y^2}{a}-b.$$

$$\text{Hence } y=\overline{\frac{a^2v+a^2c}{4}}\Big|^{\frac{1}{3}}=\overline{\frac{au-\frac{1}{2}a^2+ab}{3}}\Big|^{\frac{1}{2}}.$$

$$\text{Suppofe } c=0,\ \text{and } b=\tfrac{1}{2}a;\ \text{then } \overline{\frac{a^2v}{4}}\Big|^{\frac{1}{3}}=\overline{\frac{au}{3}}\Big|^{\frac{1}{2}},$$

$$\text{and confequently } \frac{27av^2}{16}=u^3.$$

Therefore the evolute aCd is, in this example, the femicubical Parabola: And AI being $=\frac{a}{2}$, and G coinciding with the point I, that point is the center of curvature correfponding to the vertex A of the Parabola APD.

6.

The curve APD being of the fpiral kind, whofe ordinates all iffue from the point G, fuppofe aCd to be the curve by whofe evolution APD is defcribed: join CG, CP; and draw GH, GQ perpendicular to CP and the tangent PQ refpectively. Fig. 32.

Call GP, y; GQ, (= HP,) p; the curve aCd, w; and (CP) the radius of curvature at P, R.

Then, CH being $=R-p$, and $GH=\sqrt{y^2-p^2}$, GC will be $=\sqrt{R^2-2pR+y^2}$. Now, by Corollary to Article 8. Chap. 5.

CH

$$CH : CG :: [u \perp \sqrt{R^2 - 2pR + y^2}] : [u \perp w] \; (= [u \perp R])$$

$$\text{i. e. } R - p : \sqrt{R^2 - 2pR + y^2}$$

$$:: \frac{R[u \perp R] - p[u \perp R] - R[u \perp p] + y[u \perp y]}{\sqrt{R^2 - 2pR + y^2}} : [u \perp R].$$

From whence it is evident, that

$$y[u \perp y] - R[u \perp p] \text{ is } = 0;$$

and consequently $R = \frac{y[u \perp y]}{[u \perp p]}$, u being any function of y.

Example. *Suppose* APD *to be the Logarithmic Spiral; and* s *the sine of the invariable angle* GPQ, *made by the ray* GP *and tangent* PQ, *radius being unity.*

Then GQ ($= p$) will be $= sy$, and consequently $R = \frac{y[u \perp y]}{[u \perp p]}$ $= \frac{y}{s}$.

Corollary. CP ($= R$) being $= \frac{y}{s}$, and $p = sy$; GH will be $= \sqrt{y^2 - s^2y^2} = \sqrt{1 - s^2} \times y$, and CG $= \sqrt{\frac{y^2}{s^2} - y^2} = \sqrt{1 - s^2} \times \frac{y}{s}$. Therefore the sine of the angle GCH will always be $= s$; and consequently, in this example, the evolute *aCd* is the same curve as the involute APD, but placed in a different position.

7.

Fig. 33. Suppose *opq* to be a kind of cycloid described by the point p carried about with the curve *a*B revolving along the immoveable circular arc AB, as in Art. 10. Chap. 5.

From D, the center of the immoveable circular arc, draw D*g* perpendicular to *p*B continued; and, to the continuation of DB, draw the perpendicular *pc*. Call the radius DB, R; B*p*, x; D*p*, y; B*g*, w; and let *fp*, the radius of curvature of the cycloid at p, be called E. Then the tangent *pt* being (by Chap. 5. Art. 10.) parallel to *g*D, D*t* ($= p$) the perpendicular from D on that tangent will be $= w + x$. Moreover, R being to to w as x to Bc $(=\frac{wx}{R})$, $\overline{cp}|^2$ will be $= x^2 - \frac{w^2x^2}{R^2}$, and $y^2 = \overline{cD}|^2 + \overline{cp}|^2 = \overline{R + \frac{wx}{R}}|^2 + x^2 - \frac{w^2x^2}{R^2} = R^2 + x^2 + 2wx$.

Therefore E $(= \frac{y[u \perp y]}{[u \perp p]}$ by the last Article) is equal to $\frac{x[u \perp x] + w[u \perp x] + x[u \perp w]}{[u \perp w] + [u \perp x]}$.

SCHOLIUM. If R be infinite, i. e. if AB (instead of being a curve) be a right line: Then Ag being perpendicular on pB; and Bg, Bp, Ap, and AB being called w, x, y, and z respectively; we have p (the perpendicular from A on the tangent to the cycloid at the point p) $= x - w$, and $y^2 = \overline{Ag}|^2 + \overline{pg}|^2 = z^2 - w^2 + \overline{x - w}|^2 = z^2 + x^2 - 2wx$. Therefore, in this case, E is $= \frac{[u \perp \overline{z^2 + x^2 - 2wx}]}{2.[u \perp \overline{x - w}]}$. Fig. 34.

EXAMPLE I. *Let* aB *be a circular arc whose center is* d; *and call the radius* Bd, r; dp, ϱ. Then, $x^2 - \overline{Bc}|^2$ being $= \overline{cp}|^2 = \varrho^2 - \overline{Bc - r}|^2 = \varrho^2 - r^2 + 2r \times Bc - \overline{Bc}|^2$, it is evident that Bc, or its equal $\frac{wx}{R}$, is $= \frac{x^2 + r^2 - \varrho^2}{2r}$: Hence $w = \frac{R}{2rx} \times \overline{x^2 + r^2 - \varrho^2}$. Fig. 33.

Consequently p $(= w + x)$ is $= \frac{R + 2r}{2r} . x + \frac{R . \overline{r^2 - \varrho^2}}{2r} . x^{-1}$,

$$y^2 = R^2 + \frac{R}{r} . \overline{r^2 - \varrho^2} + \frac{R + r}{r} . x^2, \text{ and}$$

$$E = \frac{2 . \overline{R + r} . x^2}{\overline{R + 2r} . x^2 - R . \overline{r^2 - \varrho^2}} = \frac{\overline{R + r} . x^2}{\overline{R + r} . x - rw}.$$

COROLLARY I. Suppose the describing point p to be in the periphery of the moveable circle. Then, ϱ being $= r$, p is $= \frac{R + 2r}{2r} . x$, $y^2 = R^2 + \frac{R + r}{r} . x^2$, and $E = \frac{2 . \overline{R + r}}{R + 2r} . x$.

From whence we have $y^2 = R^2 + p^2 - \frac{R^2}{\overline{R+2r}|^2} \cdot p^2$, and $\sqrt{E^2 - 2pE + y^2} = Df = \overset{''}{y}$ (the ray from D to the point of the evolute corresponding to the point p of the involute) $= \sqrt{R^2 - \frac{R^2 \cdot \overline{R+r}}{r \cdot \overline{R+2r}|^2} \cdot x^2}$. Now $Dg = \overset{''}{p}$ being $= \sqrt{R^2 - w^2} = \sqrt{R^2 - \frac{R^2 x^2}{4r^2}}$, x^2 is $= 4r^2 - \frac{4r^2 p^2}{R^2}$, and consequently $\overset{''}{y}{}^2 = \frac{R^4}{\overline{R+2r}|^2} + \overset{''}{p}{}^2 - \frac{R^2}{\overline{R+2r}|^2} \cdot \overset{''}{p}{}^2$.

Bringing the last expression to the same form with the value of y^2 we have $\overset{''}{y}{}^2 = \overset{''}{R}{}^2 + \overset{''}{p}{}^2 - \frac{\overset{''}{R}{}^2}{\overline{\overset{''}{R} + 2\overset{''}{r}}|^2} \cdot \overset{''}{p}{}^2$, where $\overset{''}{R}{}^2$ must be $= \frac{R^4}{\overline{R+2r}|^2}$, and $\frac{\overset{''}{R}{}^2}{\overline{\overset{''}{R}+2\overset{''}{r}}|^2} = \frac{R^2}{\overline{R+2r}|^2}$. Therefore, $\overset{''}{R}$ is $= \frac{R^2}{R+2r}$, $\overset{''}{r} = \frac{Rr}{R+2r}$, and $\overset{''}{R} + 2\overset{''}{r} = R$. Hence it is plain, that the evolute of the cycloid *opq* is another cycloid, which may be described by a point in the periphery of a circle whose radius is $\frac{Rr}{R+2r}$ revolving upon the immoveable circle whose radius is $\frac{R^2}{R+2r}$ and center D; and the position of the evolute with respect to the involute is very obvious.

Corollary II. Let the fixed point P be so situated, that AP and DP shall be respectively equal to *ap*, *dp*; and let r be supposed equal to R. Then, the arcs AB, aB being equal, if the ray PB be drawn, the angle DBP will always be equal to the angle DB*g*. Therefore the evolute of the cycloid *opq* is the caustic by reflection of the circle AB, P being the focus of the incident rays.

Now, R being $= r$, E will be $= \frac{4r^3}{3r^2 - r^2 + p^2} = \frac{2r^3}{2r - w}$, and $Bf = E - x$ (the distance of the focus of the reflected rays from the

the point of incidence) $= \frac{x^2 + \overline{r^2 - \rho^2}.x}{3x^2 - r^2 + \rho^2} = \frac{wx}{2x - w}$, where x is $=$ PB, and $\rho =$ PD.

Hence it is obvious, that if ρ be $= r$, i. e. if the focus of the incident rays be any where in the periphery of the circle AB, the focal diſtance Bf will always be $= \frac{x}{3} = \frac{2w}{3}$.

COROLLARY III. Let r and ρ be ſuppoſed each equal to $\frac{1}{2}$R. Then, the arcs AB, aB being equal, and the deſcribing point p coinciding with a, if any ray PB be drawn perpendicular to AD, the angles DBP, DBf will always be equal. Fig. 35.

Therefore the evolute of the cycloid opq is the cauſtic by reflection of the circle AB, the incident rays being parallel.

Now, $\frac{1}{2}$R being $= r = \rho$, E will be $= \frac{1}{2}x$; and the focal diſtance B$f = \frac{1}{4}x = \frac{1}{2}w$.

COROLLARY IV. AB being any reflecting curve whatever; the focus of rays reflected from any point (B) thereof may be found by the above theorems, by conſidering R as the radius, and D as the center of curvature of ſuch curve at the point of incidence.

EXAMPLE II. *Let* ap *be perpendicular to the tangent to the moveable curve at* a, *and the points* P, A, D *in a right line: and ſuppoſe the curve* aB *of ſuch a nature, that, the arc* aB *being equal to the arc* AB, pB ($= x$) *ſhall always be equal to* n × PB, n *being invariable.* Fig. 36.

Then PB being continued, and Dh drawn perpendicular thereto, $n\sqrt{R^2 - v^2}$ will be $= \sqrt{R^2 - w^2}$, v being put for Bh: for it follows from Chap. 5. Art. 8. that the ſines of the angles DBh, DBg will be to each other in the invariable ratio of 1 to n reſpectively; thoſe ſines being reſpectively equal to the coſine of the angle made by PB and the common tangent to the two curves at B, and the coſine of the angle made by pB and the ſame tangent.—Moreover, calling AP, d; $\overline{d + R}|^2 - R^2 + v^2$ will be $= \overline{v + \frac{x}{n}}|^2$.—By means of which equations

M 2 [u –

$[u \pm w]$ and $[u \pm x]$ may be eaſily expunged out of the equation $E = \frac{x[u \pm x] + w[u \pm x] + x[u \pm w]}{[u \pm w] + [u - x]}$, and then we ſhall have

$$E = \frac{\overline{w - nv} \cdot x^2 + \overline{w^2 - n^2v^2} \cdot x}{\overline{w - nv} \cdot x - n^2v^2}.$$

COROLLARY I. The ſines of the angles DBh, DBg being to each other in the invariable ratio of 1 to n reſpectively, the evolute of the curve apq is the cauſtic by refraction of the circle AB, P being the focus of the incident rays, and 1 to n the ratio of the ſine of incidence to that of refraction. Conſequently, taking x from the value of E, and writing ny inſtead of x; Bf, the diſtance of the focus of the refracted rays from the point of incidence, will be found $= \frac{w^2y}{wy - nvy - nv^2}$, where y is equal to PB.

COROLLARY II. AB being any refracting curve whatever, the focus of rays refracted at any point (B) thereof may be found by the theorem in the preceding Corollary, by conſidering R as the radius, and D as the center of curvature of ſuch curve at the point of incidence.

Fig. 34. SCHOLIUM. If R be infinite, i. e. if AB (inſtead of being a curve) be a right line, to which AP is perpendicular. Then, Ah, Ag being reſpectively perpendicular to BP, Bp; and AP, Bh, Bg, Bp, BP, and AB being called d, v, w, x, y, and z reſpectively; we have $y^2 - d^2 = z^2 = vy$, $x = ny$, and $w = nv$. From whence we get $x - w = ny - nv = ny - \frac{ny^2 - nd^2}{y} = \frac{nd^2}{y}$; and $z^2 + x^2 - 2wx = y^2 - d^2 + n^2y^2 - 2n^2vy = y^2 - n^2y^2 - d^2 + n^2d^2$.

Therefore, by the Scholium to Article 7. E is $= \frac{\overline{n^2 - 1} \cdot y^3}{nd^2}$.

Conſequently, in this caſe, by what is obſerved in Corollary 1. E − x, the diſtance of the focus of the refracted rays from the point of incidence B, is $= \frac{\overline{n^2 - 1} \cdot y^3}{nd^2} - ny$, P being the focus of the incident rays.

8. To

8.

To find the point F where any refracted ray BF meets the axis AF of any refracting curve AB. Fig. 37. Draw the incident ray PB, the ordinate BC, the tangent Bt, and BD perpendicular to that tangent. Call the abscissa AC, x; the ordinate BC, y; the curve AB, z; PB, v; BF, w; AP, d; AF, e: and let the sine of incidence be to that of refraction, as 1 to n. Then, the subtangent, ordinate, and tangent being to each other as 1, $[x \perp y]$, and $[x \perp z]$ respectively, by Chap. 5. Art. 5. we have

$1 : [x \perp y] :: y : y[x \perp y] = CD$; and (radius being 1)

$[x \perp z] : 1 :: 1 : \frac{1}{[x \perp z]}$, the sine of the angle tBC = BDC;

$[x \perp z] : 1 :: [x \perp y] : \frac{[x \perp y]}{[x \perp z]}$, the cosine of the same angle.

Moreover $v : 1 :: d + x : \frac{d+x}{v}$, the sine of the angle PBC;

and $v : 1 :: y : \frac{y}{v}$, the cosine of PBC.

Now the cosine of the difference of two angles being equal to the rectangle of the two sines *plus* the rectangle of the two cosines of those two angles, $\frac{d+x}{v[x \perp z]} + \frac{y[x \perp y]}{v[x \perp z]}$ is the cosine of the angle tBP, or the sine of the angle of incidence.

Therefore, n times this last sine being the sine of the angle of refraction DBF, we have

$$n \times \frac{d+x+y[x \perp y]}{v[x \perp z]} : \frac{1}{[x \perp z]} :: e - x - y[x \perp y] \; (= DF) : w.$$

Consequently, v being $= \sqrt{y^2 + \overline{d+x}^2}$, and $w = \sqrt{y^2 + \overline{e-x}^2}$,

$$n \times \frac{d+x+y[x \perp y]}{\sqrt{\overline{d+x}^2 + y^2}} \text{ is } = \frac{e-x-y[x \perp y]}{\sqrt{\overline{e-x}^2 + y^2}},$$

by means of which equation, and the equation of the given refracting curve, e may be readily determined.

SCHOLIUM I. If d be infinite, i. e. if the incident ray PB be parallel to the axis of the curve, the sine of PBC will be $= 1$, and

and the cosine = o. Therefore, in that case, $n : 1 :: DF : BF$, and consequently $n = \frac{e - x - y[x \perp y]}{\sqrt{\overline{e-x}^2 + y^2}}$.

SCHOLIUM II. That the refracted rays may be all parallel, the angle made by BD and any ray after refraction at B must be equal to the angle BDC. Therefore $n \times \frac{d + x + y[x \perp y]}{v[x \perp z]}$ must be $= \frac{1}{[x \perp z]}$; and consequently, in this case, $n \times \frac{d + x + y[x \perp y]}{\sqrt{\overline{d+x}^2 + y^2}}$ must be = 1.

EXAMPLE I. *The refracting curve being* DESCARTES' *Oval, whose equation is* $n\sqrt{\overline{a+x}^2 + y^2} - na = b - \sqrt{\overline{b-x}^2 + y^2}$; we, by residual division, get $n \times \frac{a + x + y[x \perp y]}{\sqrt{\overline{a+x}^2 + y^2}} = \frac{b - x - y[x \perp y]}{\sqrt{\overline{b-x}^2 + y^2}}$.

Which equation being compared with the equation $n \times \frac{d + x + y[x \perp y]}{\sqrt{\overline{d+x}^2 + y^2}} = \frac{e - x - y[x \perp y]}{\sqrt{\overline{e-x}^2 + y^2}}$ above found, it appears that the sines of incidence and refraction being as 1 to n, if (n being less than 1, and a and b positive quantities,) d be $= a$, e will always be $= b$; and the refracted rays, from every point of the curve, will all meet the axis in one and the same point, which therefore will be their common focus, without any aberration.

EXAMPLE II. *Suppose the curve* AB *to be* a circle, *whose radius is* a. Then, y^2 being $= 2ax - x^2$, $y[x \perp y]$ is $= a - x$; and by substituting these values of y^2 and $y[x \perp y]$ for their respective equals in the equation $n \times \frac{d + x + y[x \perp y]}{\sqrt{\overline{d+x}^2 + y^2}} = \frac{e - x - y[x \perp y]}{\sqrt{\overline{e-x}^2 + y^2}}$, we have $\frac{n . \overline{a+d}}{\sqrt{d^2 + 2 . \overline{a+d} . x}} = \frac{e - a}{\sqrt{e^2 + 2 . \overline{a-e} . x}}$: from whence, when a, d, n, and x are given, e may be readily found; and, consequently, the aberration of any ray from the focus of rays falling on or very near the vertex A.

From

From the last equation we get $\frac{\sqrt{d^2 + 2.\overline{a+d}.x}}{\sqrt{e^2 + 2.\overline{a-e}.x}} = \frac{n.\overline{a+d}}{e-a}$, therefore $\frac{\sqrt{b^2d^2 + 2b^2.\overline{a+d}.x}}{\sqrt{e^2 + 2.\overline{a-e}.x}}$ is $= \frac{bn.\overline{a+d}}{e-a}$, b being of any value whatever. Now supposing $e^2 = b^2d^2$, and $a - e = b^2 \times \overline{a+d}$, $\frac{bn.\overline{a+d}}{e-a}$, it is evident, must be $= 1$, let x be what it will: From which equations we have $e = bd = -nd = \overline{n+1}.a$, and $d = -\frac{n+1}{n}a$. Hence it appears, that if P be on the concave side of AB, and its distance from A be $= \frac{n+1}{n}a$, the refracted rays from every point of the curve will all diverge from one point in the axis, the distance of which point from A will be $= \overline{n+1}.a$.

And it follows, that rays converging to P, (situated as just now mentioned,) and falling on the convexity of the circle, will all be refracted to F, AF being $= \overline{n+1}.a$.

EXAMPLE III. *Let* AB *be* an ellipsis, *whose equation is* $y^2 = ax - \frac{a}{b}x^2$, *a being the parameter, and b the transverse axis.*

Then will $-\sqrt{\overline{g-x}|^2 + y^2}$ be $= -\sqrt{\overline{g-x}|^2 + ax - \frac{a}{b}x^2}$, g being of any value whatever. Hence, by residual division, we get $\frac{g - x - y[x \div y]}{\sqrt{\overline{g-x}|^2 + y^2}} = \frac{g - x - \frac{1}{2}a + \frac{a}{b}x}{\sqrt{\overline{g-x}|^2 + ax - \frac{a}{b}x^2}}$. By comparing this equation with that in Schol. 1. it appears that, the incident rays being parallel to the axis of the curve, e will always be equal to the invariable quantity g, let x be what it will, if $\frac{g - x - \frac{1}{2}a + \frac{a}{b}x}{\sqrt{\overline{g-x}|^2 + ax - \frac{a}{b}x^2}}$ be always $= n$, or $\overline{g - \frac{1}{2}a + \overline{\frac{a}{b} - 1}.x}|^2$ always $= n^2 \times \overline{\overline{g-x}|^2 + ax - \frac{a}{b}x^2}$; i. e. $\overline{g - \frac{1}{2}a}|^2 +$

$2 \cdot \overline{g - \frac{1}{2}a} \cdot \overline{\frac{a}{b} - 1} \cdot x + \overline{\frac{a}{b} - 1}\,\big|^2 \cdot x^2$ always $= n^2g^2 + n^2 \cdot \overline{a - 2g} \cdot x - n^2 \cdot \overline{\frac{a}{b} - 1} \cdot x^2$. Now, that this laſt may be a true equation, x being of any value whatever, $\overline{g - \frac{1}{2}a}\,\big|^2$ muſt be $= n^2g^2$, $2 \cdot \overline{g - \frac{1}{2}a} \cdot \overline{\frac{a}{b} - 1} = n^2 \cdot \overline{a - 2g}$ $(= - 2n^2 \cdot \overline{g - \frac{1}{2}a})$, and $\overline{\frac{a}{b} - 1}\,\big|^2 = - n^2 \cdot \overline{\frac{a}{b} - 1}$. And from theſe equations we have $\frac{a}{b} = 1 - n^2$, and $g = \frac{1}{2}a \times \frac{1}{1-n} = \frac{1}{2}b \times \overline{1 + n}$; which value of g is the diſtance of the remoter focus of the ellipſis from the vertex, when $\frac{a}{b}$ is $= 1 - n^2$. It is evident therefore, that, $\frac{a}{b}$ being $= 1 - n^2$, and the incident rays being parallel to the tranſverſe axis, and falling on the convexity of the curve, the refracted rays from every point of the figure will, without any aberration, all converge to the focus juſt now mentioned.

And (reverſing the ratio of the ſines of incidence and refraction) it is obvious that, $\frac{a}{b}$ being $= \frac{n^2 - 1}{n^2}$, and the incident rays iſſuing from one of the focuſes of the ellipſis and falling on the concavity of the farther half thereof, the rays after refraction will all proceed in a direction parallel to the tranſverſe axis.

Writing $-b$ inſtead of b, the equation of the curve becomes $y^2 = ax + \frac{a}{b}x^2$, correſponding to an hyperbola whoſe parameter is a, and tranſverſe axis b. It follows therefore, that, $\frac{a}{b}$ being $= n^2 - 1$, and the incident rays being parallel to the tranverſe axis, and falling on the concavity of the curve AB, now ſuppoſed an hyperbola, the refracted rays will all converge to the focus of the conjugate hyperbola,

And (reverſing the ratio of the ſines of incidence and refraction) it is obvious that, $\frac{a}{b}$ being $= \frac{1 - n^2}{n^2}$, and the incident rays iſſuing from the focus of the conjugate hyperbola, and falling on the convexity of the hyperbola AB, the rays, after refraction at the laſt-mentioned curve, will, from every point thereof, proceed in a direction parallel to the tranſverſe axis.

THE

Plate IV.

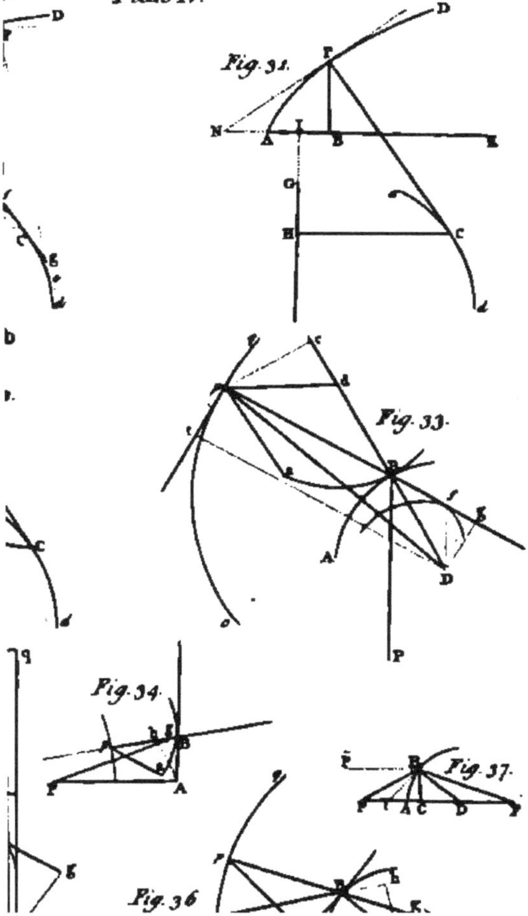

John White his Book. Bennington near Boston Lincolnshire old England. 5th Aug.t 1819.

THE

RESIDUAL ANALYSIS.

CHAP. VIII.

Of the greatest *and* least ORDINATES, *the* POINTS *of* contrary flexion *and* reflexion, *and the* double *and* triple &c. POINTS *of* curve Lines.

I.

IF, y being some function of x, y be always less or always greater than $y_{,}$, when $x_{,}$ taken between certain limits, is either less or greater than x, how near soever $x_{,}$ be taken to x; the value of y, or the quantity denoted thereby, is considered as a *maximum* or *minimum*, without regard to any value it may have when x is not taken between the said limits.

2.

The ordinate from a point of a curve is a *maximum*, or *minimum*, when, the curve being immediately continued on both sides thereof, it is greater, or less, than the ordinates which may be drawn, on either side of it, from the adjoining parts of the curve.

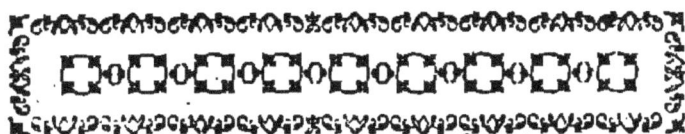

3.

Suppose y to be the ordinate of a curve corresponding to the abscissa x; and $y_{,}$ another ordinate (of the same curve) corresponding to the abscissa $x_{,}$. Then if y be a *maximum*, it will be greater than $y_{,}$, whether $x_{,}$ be less or greater than x, provided $x_{,}$ be taken between certain limits: and consequently the residual $y - y_{,}$ will, in such case, be always positive, when $x_{,}$ is taken either less or greater than x, between those limits.

But if y be a *minimum*, $y - y_{,}$ will be always negative, when $x_{,}$ is taken either less or greater than x, between certain limits.

Now seeing that, when y is a *maximum* or *minimum*, the value of the residual $y - y_{,}$ must be always positive or always negative, when $x_{,}$ is taken either less or greater than x, between certain limits: it is obvious, that, in such case, the value of $\overline{y - y_{,}} \div \overline{x - x_{,}}$ will, accordingly, be positive when $x_{,}$ is less than x, and negative when $x_{,}$ is greater than x; or positive when $x_{,}$ is greater than x, and negative when $x_{,}$ is less than x, how near soever $x_{,}$ be taken to x. Therefore, by Chap. 4. Art. 1.

$$[x \div y] \text{ will then be } = 0, \text{ or } \frac{1}{[x \div y]} = 0.$$

Moreover, supposing b to denote a value of x in either of those equations, and d the correspondent value of y, d being a *maximum* or a *minimum*, $\frac{y - d}{x - b}$ will, accordingly, be positive when x is less than b, and negative when x is greater than b; or positive when x is greater than b, and negative when x is less than b,

how

how near ſoever x be taken to b, between certain limits. Therefore, by what is ſaid in Chap. 4. Art. 3. that expreſſion, viz.

$$\frac{y-d}{x-b} \text{ will be } = \overline{x-b}|^m \times Q;$$

m being ſuppoſed an odd number, or a fraction whoſe numerator and denominator are both odd numbers; and Q ſome algebraic expreſſion conſiſting of ſuch quantities, that its value ſhall be real when x is either leſs or greater than b, (between certain limits,) and that neither it nor its reciprocal ſhall vaniſh when x is equal to b.

And it likewiſe follows from the article laſt mentioned, that, the abſciſſa being equal to b, the correſpondent ordinate ſhall be a *maximum* or a *minimum*, according as q, the value of Q when x is equal to b, is negative or poſitive, m being as juſt now ſpecified. But, if m comes out contrary to our ſuppoſition, the abſciſſa, when equal to b, will not correſpond to an ordinate that is either a *maximum* or a *minimum*.—This we ſhall, by and by, explain by proper examples.

The equation $\frac{y-d}{x-b} = \overline{x-b}|^m \times Q$ being divided by $\overline{x-b}|^m$, we have $\frac{y-d}{\overline{x-b}|^{m+1}} = Q$; where ($d$ being finite) $m+1$ muſt be poſitive, otherwiſe q and $\frac{1}{q}$ cannot be finite. Now, by Chap. 2. Cor. to Art. 6. the value of $\frac{y-d}{\overline{x-b}|^{m+1}}$, when x is $= b$, is equal to the value of the quotient of $[x \perp y]$ divided by $\overline{m+1} \, . \, \overline{x-b}|^m$, when x is taken $= b$: Therefore q will be equal to the value of $\frac{[x \perp y]}{\overline{m+1} \, . \, \overline{x-b}|^m}$, when x is taken as juſt now mentioned. Hence it is evident, that m and q may be found by reſolving $[x \perp y]$ into two ſuch factors, (F and G,) that one of them (F) ſhall be ſome power of $x-b$, and the other (G) ſhall neither vaniſh nor become infinite when x is therein taken equal to b: for by comparing $\overline{x-b}|^m$ with F, m will be known; and q will be the value of $\frac{G}{m+1}$ when x is equal to b. Moreover, Q will be real or imaginary when x is taken leſs

 or

or greater than b, between certain limits; according as G is real or imaginary when x is so taken. Therefore, by means of the said factors F and G, we may readily know whether the ordinate corresponding to the abscissa b be a *maximum* or *minimum*, without assigning the general value of Q.

EXAMPLE I. *Suppose* $y = ax - x^2$, a *being invariable.*

Then we shall have $[x \perp y] = a - 2x = \overline{x - \frac{a}{2}} \times -2 = 0$; where x is $= \frac{a}{2}$.

Now, b being $= \frac{a}{2}$, we have $F = x - \frac{a}{2} = \overline{x - b}|^m$, and $G = -2$. Therefore, m being $= 1$, and q (the value of $\frac{G}{m+1}$ when x is $= b$) being a negative quantity, y is a *maximum* when x is $= \frac{a}{2}$.

EXAMPLE II. *Suppose* $y = x^4 - a^3 x$, *where* a *is invariable.*

Then we have

$$[x \perp y] = 4x^3 - a^3 = \overline{x - \frac{a}{4^{\frac{1}{3}}}} \times \overline{4^{\frac{1}{3}}a^2 + 4^{\frac{2}{3}}ax + 4x^2} = 0;$$

where the real value of x is $\frac{a}{4^{\frac{1}{3}}}$.

Now, b being $= \frac{a}{4^{\frac{1}{3}}}$, we have $F = x - \frac{a}{4^{\frac{1}{3}}} = \overline{x - b}|^m$, and $G = 4^{\frac{1}{3}}a^2 + 4^{\frac{2}{3}}ax + 4x^2$. Therefore, m being $= 1$, and q (the value of $\frac{G}{m+1}$ when x is $= b$) being a positive quantity, y is a *minimum* when x is $= \frac{a}{4^{\frac{1}{3}}}$.

EXAMPLE III. *Suppose* $y = x^2 + \overline{a^2 - x^2}|^{\frac{3}{2}}$, *where* a *is an invariable positive quantity.*

Then we shall have

$$[x \perp y] = \frac{2x \times \overline{a^2 - x^2}|^{\frac{1}{2}} - 2x^3}{\overline{a^2 - x^2}|^{\frac{1}{2}}} \left(= \frac{8a^2x - 16x^4}{\overline{a^2 - x^2}|^{\frac{1}{2}} \times \overline{g^2x^{-2} + 6g + 12x^2}}\right.^{*} =$$

* $2x \times \overline{a^2 - x^2}|^{\frac{1}{2}} - 2x^3$ being supposed $= g$, $2x \times \overline{a^2 - x^2}|^{\frac{1}{2}}$ will be $= g + 2x^3$, and

$= \dfrac{-8x \times \overline{x - \frac{a}{2^{\frac{1}{3}}}} \times \overline{2^{\frac{2}{3}}a^2 + 2^{\frac{1}{3}}ax + 2x^2}}{\overline{a^2 - x^2}|^{\frac{1}{3}} \times \overline{g^2x^{-2} + 6g + 12x^2}}$, g being put for the numerator $2x \times \overline{a^2 - x^2}|^{\frac{1}{3}} - 2x^2) = 0$; and $\dfrac{1}{[x \perp y]} = \dfrac{\overline{a^2 - x^2}|^{\frac{2}{3}}}{2x \times \overline{a^2 - x^2}|^{\frac{1}{3}} - 2x^2} = \dfrac{\overline{x - a}|^{\frac{2}{3}} \times \overline{a^2 + ax + x^2}|^{\frac{2}{3}}}{2x^2 - 2x \times \overline{a^2 - x^2}|^{\frac{1}{3}}} = 0.$

In the equation $[x \perp y] = 0$, the real values of x are 0 and $\frac{a}{2^{\frac{1}{3}}}$; and, in the equation $\frac{1}{[x \perp y]} = 0$, the real value of x is a.

Taking b equal to 0, we have $F = x - 0 = \overline{x - b}|^m$, and $G = \dfrac{-8 \times \overline{x - \frac{a}{2^{\frac{1}{3}}}} \times \overline{2^{\frac{2}{3}}a^2 + 2^{\frac{1}{3}}ax + 2x^2}}{\overline{a^2 - x^2}|^{\frac{1}{3}} \times \overline{g^2x^{-2} + 6g + 12x^2}}$. Therefore, m being manifestly $= 1$, and q (the value of $\frac{G}{m + 1}$ when x is $= b$) being a positive quantity; y is a *minimum* when x is $= 0$.

Taking b equal to $\frac{a}{2^{\frac{1}{3}}}$, we have $F = x - \frac{a}{2^{\frac{1}{3}}} = \overline{x - b}|^m$, and $G = \dfrac{-8x \times \overline{2^{\frac{2}{3}}a^2 + 2^{\frac{1}{3}}ax + 2x^2}}{\overline{a^2 - x^2}|^{\frac{1}{3}} \times \overline{g^2x^{-2} + 6g + 12x^2}}$. Therefore, m being $= 1$, and q being a negative quantity, y is a *maximum* when x is $= \frac{a}{2^{\frac{1}{3}}}$.

Taking b equal to a, we have $F = \overline{x - a}|^{-\frac{2}{3}} = \overline{x - b}|^m$, and $G = \dfrac{2x^2 - 2x \times \overline{a^2 - x^2}|^{\frac{1}{3}}}{\overline{a^2 + ax + x^2}|^{\frac{2}{3}}}$. Therefore m being $= -\frac{2}{3}$, and q being a positive quantity, y is a *minimum* when x is $= a$.

---

and $8a^2x^4 - 16x^6 = g^3 + 6g^2x^2 + 12gx^4$. From whence it is evident, that if g be multiplied by $g^2 + 6gx^2 + 12x^4$, the product will be equal to $8a^2x^4 - 16x^6$; and therefore it is plain that $\dfrac{2x \times \overline{a^2 - x^2}|^{\frac{1}{3}} - 2x^2}{\overline{a^2 - x^2}|^{\frac{1}{3}}}$ is equal to $\dfrac{8a^2x^3 - 16x^5}{\overline{a^2 - x^2}|^{\frac{1}{3}} \times \overline{g^2 + 6gx^2 + 12x^4}} = \dfrac{8a^2x - 16x^3}{\overline{a^2 - x^2}|^{\frac{1}{3}} \times \overline{g^2x^{-2} + 6g + 12x^2}}$, the numerator and denominator being each divided by x^2, the variable quantity by which both $8a^2x^3 - 16x^5$ and $g^2 + 6gx^2 + 12x^4$ are divisible.

EXAMPLE

Example IV. *Suppose* $y = a^4x - a^3x^2 - \frac{2}{3}a^2x^3 + \frac{3}{2}ax^4 - \frac{3}{5}x^5$, *where* a *is an invariable positive quantity.*

Then will $[x \div y]$ be $= a^4 - 2a^3x - 2a^2x^2 + 6ax^3 - 3x^4 =$

$$- 3 \times \overline{x - \frac{a}{\sqrt{3}}} \times \overline{x + \frac{a}{\sqrt{3}}} \times \overline{x - a}|^2 = 0,$$

where x is equal to $\frac{a}{\sqrt{3}}$, or $-\frac{a}{\sqrt{3}}$, or a.

Taking b equal to $\frac{a}{\sqrt{3}}$, we have $F = x - \frac{a}{\sqrt{3}} = \overline{x - b}|^m$, and $G = -3 \times \overline{x + \frac{a}{\sqrt{3}}} \times \overline{x - a}|^2$. Therefore, m being evidently $= 1$, and q (the value of $\frac{G}{m+1}$ when x is $= b$) being a negative quantity, y is a *maximum* when x is $= \frac{a}{\sqrt{3}}$.

Taking b equal to $-\frac{a}{\sqrt{3}}$, we have $F = x + \frac{a}{\sqrt{3}} = \overline{x - b}|^m$, and $G = -3 \times \overline{x - \frac{a}{\sqrt{3}}} \times \overline{x - a}|^2$. Therefore, m being $= 1$, and q being a positive quantity, y is a *minimum* when x is $= -\frac{a}{\sqrt{3}}$.

Taking b equal to a, we have $F = \overline{x - a}|^2 = \overline{x - b}|^m$, and $G = -3 \times \overline{x - \frac{a}{\sqrt{3}}} \times \overline{x + \frac{a}{\sqrt{3}}}$. Therefore m is $= 2$; which being contrary to our supposition, it follows, that y is neither a *maximum* nor a *minimum* when x is $= a$, notwithstanding this value of x is determined in the same manner as are the other two values above specified.

Example V. *Suppose* $y = x + \overline{a^2 - x^2}|^{\frac{1}{2}}$, *where* a *is an invariable positive quantity.*

Then we have $[x \div y] = \frac{\overline{a^2 - x^2}|^{\frac{1}{2}} - x}{\overline{a^2 - x^2}|^{\frac{1}{2}}} (= \frac{a^4 - 2a^2x^2}{\overline{a^2 - x^2}|^{\frac{1}{2}} \times \overline{a^4 + 3a^2x^2 + 3x^4}}$

$=$

$$= \frac{\overline{x - \frac{a}{2^{\frac{1}{3}}}} \times \overline{-2^{\frac{1}{3}}a^{3} - 2^{\frac{2}{3}}a^{2}x - 2a^{2}x^{2}}}{\overline{a^{3} - x^{3}}|^{\frac{2}{3}} \times \overline{g^{2} + 3gx + 3x^{2}}} \text{ (} g \text{ being put for } \overline{a^{3} - x^{3}}|^{\frac{1}{3}} - x) = 0;$$

and $\frac{1}{[x \perp y]} = \frac{\overline{a^{3} - x^{3}}|^{\frac{2}{3}}}{\overline{a^{3} - x^{3}}|^{\frac{1}{3}} - x} = \frac{\overline{x - a}|^{\frac{2}{3}} \times \overline{a^{2} + ax + x^{2}}|^{\frac{2}{3}}}{\overline{a^{3} - x^{3}}|^{\frac{1}{3}} - x} = 0.$

In the equation $[x \perp y] = 0$, the real value of x is $\frac{a}{2^{\frac{1}{3}}}$; and in the equation $\frac{1}{[x \perp y]} = 0$, the real value of x is a.

Taking b equal to $\frac{a}{2^{\frac{1}{3}}}$, we have $F = x - \frac{a}{2^{\frac{1}{3}}} = \overline{x - b}|^{m}$, and $G = \frac{-2^{\frac{1}{3}}a^{3} - 2^{\frac{2}{3}}a^{2}x - 2a^{2}x^{2}}{\overline{a^{3} - x^{3}}|^{\frac{2}{3}} \times \overline{g^{2} + 3gx + 3x^{2}}}$. Therefore, m being $= 1$, and q (the value of $\frac{G}{m + 1}$ when x is $= b$) being a negative quantity, y is a *maximum* when x is $= \frac{a}{2^{\frac{1}{3}}}$.

Taking b equal to a, we have $F = \overline{x - a}|^{-\frac{2}{3}} = \overline{x - b}|^{m}$, and $G = \frac{\overline{a^{3} - x^{3}}|^{\frac{1}{3}} - x}{\overline{a^{2} + ax + x^{2}}|^{\frac{2}{3}}}$. Therefore m is $= -\frac{2}{3}$; which being contrary to our supposition, it follows, that y is neither a *maximum* nor a *minimum* when x is $= a$.

EXAMPLE VI. *Suppose* $y = a - \frac{\overline{a - x}|^{\frac{5}{3}}}{a^{\frac{2}{3}}} + \frac{\overline{a - x}|^{2}}{a}$, *where* a *is invariable and positive.*

Then will

$$[x \perp y] \text{ be} = \frac{5 \cdot \overline{a - x}|^{\frac{2}{3}}}{3a^{\frac{2}{3}}} - \frac{2 \cdot \overline{a - x}}{a} = \frac{25 \times \overline{x - \frac{9a}{25}} \times \overline{x - a}}{3a^{\frac{2}{3}} \times \overline{5 \cdot \overline{a - x}|^{\frac{1}{3}} + 4a^{\frac{1}{3}}}} = 0;$$

where x is equal to $\frac{9a}{25}$, or to a.

Taking b equal to $\frac{9a}{25}$, we have $F = x - \frac{9a}{25} = \overline{x - b}|^{m}$, and

and $G = \frac{25 \times \overline{x-a}}{2a^{\frac{1}{2}} \times 5 \cdot \overline{a-x}|^{\frac{1}{2}} + 4a^{\frac{1}{2}}}$. Therefore, m being $= 1$, and q (the value of $\frac{G}{m+1}$ when x is $= b$) being a negative quantity, y is a *maximum* when x is $= \frac{9a}{25}$.

Taking b equal to a, we have $F = x - a = \overline{x-b}|^{m}$, and $G = \frac{25x - 9a}{2a^{\frac{1}{2}} \times 5 \cdot \overline{a-x}|^{\frac{1}{2}} + 4a^{\frac{1}{2}}}$. Here m, which is manifestly $= 1$, is agreeable to our supposition; but, the value of G being imaginary when x is greater than a, it appears, (as it likewise does from the equation of the curve,) that, when x is equal to a, the curve is not continued on both sides of the ordinate; therefore y is not then a *maximum* or *minimum*, within the meaning of our explication in Art. 1. and 2.

It may be worth while to enquire concerning the point of the curve to which the ordinate d (determined as above) corresponds, when m comes out contrary to our supposition, or G is not real, both when x is greater and when less than b: and that I purpose to do, in the next article.

4.

Fig. 38. We have found, in Chap. 5. that $\frac{y}{[x \perp y]}$ is the general value
39.
40. of the subtangent; therefore $[x \perp y]$ is the quotient of the
41. ordinate divided by the subtangent. Now it is obvious, that the
42.
43. said quotient will vanish, when, the ordinate being finite, the
44.
45. tangent is parallel to the base; and that $\left(\frac{1}{[x \perp y]}\right)$ the reciprocal of
46. the same quotient will vanish when the tangent coincides with
47.
48. the ordinate. When the tangent is parallel to the base, the
49. ordinate may be a *maximum* or *minimum*, as in Fig. 38, 39. or it may pass through a point of contrary flexion, as in Fig. 40. or correspond to a cuspid, as in Fig. 41, 42, 43. Also, when the tangent coincides with the ordinate, such ordinate may be a *maximum* or *minimum*, as in Fig. 44, 45. or it may meet the curve in a point of contrary flexure, as in Fig. 46. or correspond to

to a cuſpid, as in Fig. 47, 48. or touch a continued arch, as in Fig. 49. It appears therefore, that, b being a value of x in the equation $[x \perp y] = o$, or $\frac{1}{[x \perp y]} = o$, and d the correſpondent value of y, (as ſuppoſed above,) the ordinate d may be a *maximum* or *minimum*, or correſpond to a point of contrary flexion or reflexion, or touch a continued arch: moreover, the tangent to the correſpondent point of the curve will be parallel to the baſe, or coincide with the ordinate, according as b is determined from the equation $[x \perp y] = o$, or $\frac{1}{[x \perp y]} = o$; i. e. according as m, in the equation $y - d = \overline{x - b}|^{m+1} \times Q$, is poſitive or negative.

If the ordinate d correſponds to a point of contrary flexure, $\overline{x - b}|^{m+1} \times Q$, the value of $y - d$, muſt be negative when x is leſs than b, and poſitive when x is greater than b; or negative when x is greater than b, and poſitive when x is leſs than b. Therefore $\overline{x - b}|^{m+1}$ muſt, in that caſe, be negative or poſitive, according as x is leſs or greater than b; and conſequently $m + 1$ muſt be an odd number, or a fraction whoſe numerator and denominator are both odd numbers. From whence it follows, that m muſt then be an even number, or a fraction whoſe numerator is an even number and denominator an odd number. It is evident therefore, that, when m is as juſt now ſpecified, and G is real both when x is greater and when leſs than b, the ordinate d meets the curve in a point of contrary flexure; and the tangent at that point will be parallel to the baſe, or coincide with the ordinate, according as m is poſitive or negative: moreover, if x be increaſed after being equal to b, y will, at the ſame time, increaſe or decreaſe, according as q is poſitive or negative.

The celebrated Marquis DE L'HOSPITAL *diſtinguiſhes cuſpids (or points of reflexion,) into two kinds: when the branches of the curve which form the cuſpid have their convexity towards each other, the cuſpid is ſaid to be of the* firſt kind; *but of the* ſecond kind, *when the convexity of one of thoſe branches is towards the concavity of the other.*—In what follows, we will obſerve the ſame diſtinction.

Fig. 50. Let AE be the base upon which the abſciſſa x is meaſured when y is conſidered as the ordinate, and let AF be drawn parallel to the ſaid ordinate. Then, y being ſuppoſed to denote any abſciſſa AD, meaſured on AF; and x the correſponding ordinate DP; we have, by the preceding article, $[y \perp x] = o$, or $\frac{1}{[y \perp x]} = o$, when x is a *maximum* or *minimum*. Therefore, $[x \perp y]$ (by Art. 9. Chap. 2.) being $= \frac{1}{[y \perp x]}$, $\frac{1}{[x \perp y]}$ will be $= o$, or $[x \perp y] = o$, when x is as juſt now mentioned. Moreover, from the equation $\frac{y-d}{x-b} = \overline{x-b}|^{m} \times Q$, we get $\frac{x-b}{y-d} = \overline{y-d}|^{\frac{-m}{m+1}} \times \frac{1}{Q^{\frac{1}{m+1}}}$. Therefore, by the laſt article, $\frac{-m}{m+1}$ muſt be an odd number, or a fraction whoſe numerator and denominator are both odd numbers, when x is a *maximum* or *minimum*: Hence it follows, that m muſt then be a fraction whoſe numerator is an odd number, and denominator an even number.—Now, when DP $(= x)$ is a *maximum* or *minimum*, CP (parallel and equal to AD $= y$) either touches a continued arch; or correſponds to a cuſpid of the firſt kind, with the tangent at the point of reflexion parallel to the baſe.—It is manifeſt therefore, that, if, when DP is equal to b, the curve be continued on both ſides of that line, m muſt be as juſt now ſpecified; and the ordinate d will correſpond to ſuch a cuſpid as we laſt mentioned, or touch a continued arch, according as m is poſitive or negative.

From what has been ſaid it is evident, that, m being any number or fraction whatever, if the curve be not continued on both ſides of the ordinate d, that ordinate will correſpond to a cuſpid of the ſecond kind; except when m is a fraction whoſe numerator is an odd number, and denominator an even number, in which caſe the ſaid ordinate may correſpond to a cuſpid of either kind, or touch a continued arch.

Corollary I. The value of $[x \perp y]$ being expreſſed by a fraction, if, by ſuppoſing the numerator thereof $= o$, x be found equal to (any quantity) b; and, by ſuppoſing the denominator $= o$, x be likewiſe found equal to b; it may, in ſuch caſe, happen,

happen, that m ſhall be $= o$. When it ſo happens, the ordinate correſponding to the abſciſſa b will meet the curve in a double or triple &c. point, according as q has two or three &c. values. If, m being $= o$, q has two equal values, the ſaid ordinate will correſpond to a point of the curve where two branches thereof touch each other, either forming a cuſpid there, or being continued on both ſides of ſuch ordinate: and, if the values of q are all imaginary, ſuch ordinate will correſpond to a conjugate point.

The poſition of the tangents to the ſeveral branches of the curve, at ſuch double or triple &c. point, will be known by means of the ſeveral values of q: for, m being $= o$, q is the value of $[x \perp y]$ when x is equal to b; and therefore q will then be to unity, as the ordinate to the ſubtangent.

COROLLARY II. m being a poſitive odd number, or a poſitive fraction whoſe numerator and denominator are both odd numbers, as many ſingle different real values as q has, ſo many different branches of the curve are continued on both ſides of the ordinate d, touching each other in P, the point to which the ſaid ordinate correſponds: and two equal values of q denote two continued arches of equal curvature at P, and both turned one way; or a cuſpid of the ſecond kind, at that point.

The tangent, at P, to ſuch cuſpid, or continued arches, will be parallel to the baſe; and, according as q is negative or poſitive, the branch to which it relates will, at P, have its convexity upwards or downwards.

Imaginary values of q indicate a conjugate point at P; m being as in this, or the next Corollary.

COROLLARY III. m being a negative odd number, or a negative fraction whoſe numerator and denominator are both odd numbers, as many ſingle different real values as q has, ſo many different branches of the curve are continued on both ſides of the ordinate d, each forming a cuſpid of the firſt kind, at P: and two equal values of q denote two ſuch cuſpids, both pointing one way; or a cuſpid of the ſecond kind, at that point.

The tangent to each of theſe cuſpids will coincide with the ordinate; and, according as q is negative or poſitive, the cuſpid to which it relates will point upwards or downwards.

Corollary IV. m being an even number, or a fraction whoſe numerator is even and denominator odd, as many ſingle different real values as q has, ſo many different branches of the curve are continued on both ſides of the ordinate d, each having a point of contrary flexure at P: and two equal values of q denote two ſuch branches, both turned alike; or a cuſpid of the ſecond kind, at that point.

The tangent to ſuch cuſpid, or point of contrary flexure, will be parallel to the baſe, or coincident with the ordinate, according as m is poſitive or negative; and, according as q is poſitive or negative, y will be greater or leſs than d, when x is taken greater than b.

Corollary V. m being a poſitive fraction whoſe numerator is an odd number and denominator an even number, as many ſingle different poſitive real values as q and $\overline{-1}\rceil^{m} \times q$ have, ſo many cuſpids of the firſt kind will be formed at P: and two equal poſitive real values of q, or of $\overline{-1}\rceil^{m} \times q$, denote two ſuch cuſpids, at that point, both turned one way; or a cuſpid of the ſecond kind there.

The tangents to ſuch cuſpids will be parallel to the baſe; and, (x being ſuppoſed poſitive when the abſciſſa is on the right of the point where it begins,) according as the value of q or $\overline{-1}\rceil^{m} \times q$ is real, the branch or branches forming the cuſpid, to which ſuch value relates, will be on the right or left of the ordinate d.

m being as in this, or the following Corollary, a conjugate point is indicated, if neither q nor $\overline{-1}\rceil^{m} \times q$ be real.

Corollary VI. m being a negative fraction whoſe numerator is odd and denominator even, as many ſingle different poſitive real values as q and $\overline{-1}\rceil^{m} \times q$ have, ſo many continued arches will be touched by the ordinate d: and two equal poſitive real values of q, or of $\overline{-1}\rceil^{m} \times q$, denote, that d is a tangent to a cuſpid of the ſecond kind, at P; or that d touches two continued arches of equal curvature at that point, and both turned one way.

Moreover,

Moreover, (x being ſuppoſed poſitive when the abſciſſa is on the right of the point where it begins,) according as the value of q or $\overline{-1}|^{m} \times q$ is real, the continued arch, or the branches forming the cuſpid, to which ſuch value relates, will be on the right or left of the ſaid ordinate d.

SCHOLIUM. m being any number or fraction whatever, if q has two equal values, P, inſtead of being as above ſpecified, will ſometimes be a conjugate point. And, from this obſervation and what is ſaid above, the conſequence is obvious, when, m being of any value whatever, q has three or more equal values.

Theſe Corollaries are of great uſe, for preventing miſtakes concerning the greateſt and leaſt ordinates, and for aſcertaining the true form of a curve from the equation thereof; as will farther appear by the examples in the following articles.

5.

If the value of y be not expreſſed in terms of x, (as is frequently the caſe,) the value of $[x \perp y]$ will be expreſſed in terms containing both x and y. In which caſe, m and q will not be found as when $[x \perp y]$ is a function of x without y being concerned therein. But they may then be found as follows.

Suppoſing b to be a value of x, and d the correſpondent value of y, when either the numerator or denominator of the value of $[x \perp y]$ is $= o$; ſubſtitute $d + \overline{x-b}|^{m+1} \times Q$ for y, in the equation of the curve; (the terms being all on one ſide, and conſequently $= o$;) and, in the expreſſion which reſults upon ſuch ſubſtitution, take m of ſuch a value, (greater than -1,) that the ſaid expreſſion being divided by the loweſt power of $x-b$ in the ſeveral members thereof, the quotient ſhall not, upon taking x equal to b, conſiſt of leſs than two members, all the terms in which Q is not concerned being accounted but one member. Then, ſuch quotient being $= o$, by writing therein q inſtead of Q, and the value of b inſtead of x, q will from thence be determined.

m and q may alſo be found by ſubſtituting $d + \overline{x-b}|^{m+1} \times q$ for y in the value of $[x \perp y]$. For the reſulting expreſſion being divided

divided by $\overline{m+1} \times \overline{x-b}|^{m}$, the quotient, when x is therein taken equal to b, will, by what is proved above, be equal to q: i. e. if R be put for the expreſſion which is obtained by ſubſtituting, as juſt now mentioned, in the value of $[x \perp y]$,

$$\frac{R}{\overline{x-b}|^{m}}, \text{ when } x \text{ is } = b, \text{ will be } = \overline{m+1} \times q:$$

from whence m and q may be readily determined, by taking m of ſuch a value, that the numerator and denominatcr of the value of the expreſſion $\frac{R}{\overline{x-b}|^{m}}$ ſhall each be diviſible by one and the ſame power of $x-b$; and, when each is divided by ſuch power, neither of the quotients ſhall vaniſh upon taking x equal to b.

Remark. The exponents of the ſeveral powers of x and y being poſitive integers, the concluſion will be exactly the ſame, (and the proceſs more conciſe,) if, when the numerator of the value of $[x \perp y]$ vaniſhes, and the denominator, at the ſame time, does not vaniſh, we write b for x in the denominator, and only d for y in both numerator and denominator: Alſo, if, when the ſaid denominator vaniſhes, and the numerator, at the ſame time, does not vaniſh, we firſt write b for x in both numerator and denominator, and then ſubſtitute $d + \overline{x-b}|^{m+1} \times q$ for y in the denominator, and only d for y in the numerator; omitting (in this latter caſe) in the denominator, every term wherein y is not concerned before ſubſtitution, and every term but thoſe wherein the loweſt power of q is concerned after ſubſtitution.

For the terms omitted in the value of R in conſequence of theſe rules would, if they were therein written, always vaniſh in carrying on the proceſs; as will eaſily appear, upon conſidering what we juſt now ſaid concerning the expreſſion $\frac{R}{\overline{x-b}|^{m}}$.

Example I. *Let the equation of the curve be* $x^4 - axy^2 + cy^3 = 0$, a *and* c *being invariable poſitive quantities.*

Then, by reſidual diviſion, we have $4x^3 - ay^2 - 2axy[x \perp y] + 3cy^2[x \perp y] = 0$; and conſequently

$$[x \perp y] = \frac{4x^3 - ay^2}{2axy - 3cy^2}.$$

Suppoſing

Supposing the numerator $4x^3 - ay^2 = 0$, we, from thence and the equation of the curve, find $x = 0$, also $x = \frac{3^2a^3}{4^3c^2}$; and the respective values of y are 0, and $\frac{3^3a^4}{4^4c^3}$.

Supposing the denominator $2axy - 3cy^2 = 0$, we, from hence and the given equation, find $x = 0$, also $x = \frac{2^2a^3}{3^3c^2}$; and the respective values of y are 0, and $\frac{2^3a^4}{3^4c^3}$.

1st. Taking $b = 0$, d will be $= 0$, and $d + \overline{x - b}|^{m+1} \times Q$ (the quantity to be substituted for y, in the equation of the curve,) $= Qx^{m+1}$. Therefore $x^4 - ax^{2m+3}Q^2 + cx^{3m+3}Q^3$ will be $= 0$; where, it is obvious, that, to determine q as taught above, m may be $= 0$, also $m = \frac{1}{2}$.

If m be $= 0$, $x^4 - ax^3Q^2 + cx^3Q^3$ will be $= 0$. Therefore, in this case, (dividing by x^3, and then taking $x = 0$,) $-aq^2 + cq^3$ is $= 0$, or $q^3 - \frac{a}{c}q^2 = 0$. Here q has three values, two of which are each $= 0$, and the third $= \frac{a}{c}$.

Therefore, by Cor. 1. of the last Art. the point (A) where x begins is a triple point in the curve; and two of its branches touch each other at that point. Moreover, q being to unity, as the ordinate to the subtangent; and the value of q which relates to the said two branches being $= 0$; the tangent to these branches at A coincides with the base AF: and, the value of q which relates to the third branch being $\frac{a}{c}$, the tangent (AG) to this branch at A makes an angle GAH with the base, such, that GH (which is supposed parallel to the ordinates of the curve) is to AH, as $\frac{a}{c}$ to 1. Fig. 51.

If m be equal to $\frac{1}{2}$, $x^4 - ax^4Q^2 + cx^{\frac{9}{2}}Q^3$ will be $= 0$.

Hence

Hence $1 - aq^2 = 0$, and $q = \frac{1}{\sqrt{a}}$. Therefore, by Cor. 5. of the laſt Article, there is a cuſpid of the firſt kind at A, the tangent to which is coincident with the baſe; and (x being always ſuppoſed poſitive when the abſciſſa is on the right of the point at which it begins,) the branches forming ſuch cuſpid are on the right of the ſame point: which agrees with what we diſcovered by conſidering m equal to o.

2dly. Taking b equal to $\frac{3^2a^3}{4^3c^2}$, d will be $= \frac{3^3a^4}{4^4c^3}$. Therefore $\frac{R}{\overline{x-b}|^m}$ (abridged agreeable to our Remark) is $= \frac{4}{\overline{x-b}|^m} \times \frac{x^3 - \frac{3^6a^9}{4^9c^6}}{2abd - 3cd^2} = \frac{4}{\overline{x-b}|^m} \times \frac{x^3 - b^3}{2abd - 3cd^2} = \frac{4}{\overline{x-b}|^m} \times \frac{\overline{x-b} \times \overline{x^2 + bx + b^2}}{2abd - 3cd^2}$ $= \overline{m+1} \times q$ when x is $= b$. Whence it is manifeſt, that m is $= 1$, and $q = \frac{6b^2}{2abd - 3cd^2}$. Therefore, q being a negative quantity, it follows from what is ſaid above, that the ordinate BP ($= d$) is a *maximum* when (AB) the correſponding abſciſſa is equal to b.

3dly. Taking b equal to $\frac{2^2a^2}{3^2c^2}$, d will be $= \frac{2^3a^4}{3^4c^3}$. Therefore $\frac{R}{\overline{x-b}|}$ (abridged according to our Remark) is $= \overline{x-b}|^{-m} \times \frac{4b^3 - ad^2}{\overline{2ab - 6cd} \cdot q \cdot \overline{x-b}|^{m+1}} = \overline{m+1} \times q$ when x is $= b$. It appears therefore, that m is $= -\frac{1}{2}$, and $\frac{4b^3 - ad^2}{\overline{2ab - 6cd} \times q} = \frac{1}{2}q$: Hence q is found $= \sqrt{\frac{4b^3 - ad^2}{ab - 3cd}}$.

Now, $4b^3$ being greater than ad^2, and ab leſs than $3cd$, the value of q is imaginary, and the value of $\overline{-1}|^m \times q$ is real. Therefore, by Cor. 6. of the laſt Article, the ordinate ef ($= d$) touches a continued arch when (Ae) the correſponding abſciſſa is equal to b; and the arch ſo touched is on the left of the ſaid ordinate.

EXAMPLE

EXAMPLE II. *Let the equation of the curve be* $x^4 - 2x^2y - 4xy^2 + y^2 - y^3 = 0$.

Then, by reſidual diviſion, we have $4x^3 - 4xy - 4y^2 - 2x^2\,[x \perp y] - 8xy\,[x \perp y] + 2y\,[x \perp y] - 3y^2\,[x \perp y] = 0$; and conſequently $[x \perp y] = \frac{4x^3 - 4xy - 4y^2}{2x^2 + 8xy - 2y + 3y^2}$.

Suppoſing the numerator $4x^3 - 4xy - 4y^2 = 0$, we, from thence and the equation of the curve, find $x = 0$, alſo $x = 2$; and the reſpective values of y are 0, and -4; y being, in ſuch caſe, $= \frac{x^4 - 2x^3}{1 - x + x^2}$.

Suppoſing the denominator $2x^2 + 8xy - 2y + 3y^2 = 0$, we, from hence and the given equation, find $x = 0$, or $x = 2$, or $x = \frac{4}{27}$; and the reſpective values of y are 0, -4, and $\frac{16}{81}$; y being, in this caſe, $= \frac{9x^4 + 8x^3 - 2x^2}{16x - 20x^2 - 2}$.

1ſt. Taking b equal to 0, d will be $= 0$, and $d + \overline{x - b}|^{m+1} \times Q = Qx^{m+1}$. Therefore, this laſt quantity being ſubſtituted inſtead of y in the equation of the curve, we have

$$x^4 - 2Qx^{m+3} - 4Q^2x^{2m+3} + Q^2x^{2m+2} - Q^3x^{3m+3} = 0:$$

And here, to determine q, m muſt be $= 1$.

Conſequently $1 - 2Q - 4Q^2x + Q^2 - Q^3x^2$ is $= 0$; from whence we have $1 - 2q + q^2 = 0$: And q having two equal values, each $= 1$, there is, agreeable to Corol. 2. of the laſt Article, a cuſpid of the ſecond kind, at the point A, * where Fig. 52. x be-

* By Corollary 2. of the laſt Article, there is either a cuſpid of the ſecond kind, at the point A; or the baſe touches two continued arches, at that point. To know whether the curve has a cuſpid there, or not, ſome farther enquiry concerning the value of Q is requiſite.—Having found, that Q is $= 1$ when x is $= 0$, we may ſuppoſe $Q = 1 + x^n\ddot{Q}$, n being ſome poſitive number or fraction, and $\ddot{Q}$ ſuch a function of x, that neither it nor its reciprocal ſhall vaniſh upon taking x equal to 0.

Let therefore $1 + x^n\ddot{Q}$ be ſubſtituted inſtead of Q in the equation $1 - 2Q + Q^2 - 4Q^2x - Q^3x^2 = 0$: And then, from the reſulting equation, n may be

x begins; the tangent to which coincides with the baſe, and the convexity of the branches forming ſuch cuſpid is downwards.

2dly. Taking b equal to 2, d will be $= -4$, and $x^4 + 8x^3 - 64x + 80 - 2 \times \overline{x^2 - 16x + 28} \times \overline{x - 2}|^{m+1} \times Q + \overline{13 - 4x} \times \overline{x - 2}|^{2m+2} \times Q^2 - \overline{x - 2}|^{3m+3} \times Q^3 = 0$. It appears therefore, that $\overline{x^2 + 4x + 20} \times \overline{x - 2}|^2 + \overline{28 - 2x} \times \overline{x - 2}|^{m+2} \times Q + \overline{13 - 4x} \times \overline{x - 2}|^{2m+2} \times Q^2 - \overline{x - 2}|^{3m+3} \times Q^3$ is $= 0$; and that, to determine q, m muſt be $= 0$. Conſequently (dividing by $\overline{x - 2}|^2$, and afterwards taking $x = 2$, and $Q = q$.) $32 + 24q + 5q^2$ is $= 0$, and $q = -\frac{12}{5} \pm \frac{4}{5}\sqrt{-1}$. Here, theſe values of q being imaginary, the ordinate d, by Cor. 1. of the laſt Art. correſponds to a conjugate point, when the abſciſſa is $= 2$.

3dly. Taking b equal to $\frac{4}{27}$, d will be $= \frac{16}{81}$. Therefore $\frac{R}{\overline{x - b}|^n}$ (abridged agreeable to our Remark) is $= \overline{x - b}|^m \times \frac{4b^2 - 4bd - 4d^2}{\frac{10}{27}q \cdot \overline{x - b}|^{m+1}} = \overline{m + 1} \times q$ when x is $= b$. It is evident therefore, that m is $= -\frac{1}{2}$, and $\frac{4b^2 - 4bd - 4d^2}{\frac{10}{27}q} = \frac{1}{2}q$: Hence q is found $= \sqrt{\frac{108}{5} \times \overline{b^2 - bd - d^2}}$.

---

be determined, and alſo $\tilde{q}$, the value of $\tilde{Q}$ when x is $= 0$, in the ſame manner as m and q are determined above.

By proceeding in that manner, n is found $= \frac{1}{2}$, and $\tilde{q} = \pm 2$. Conſequently y $(= Qx^2)$ is $= x^2 + x^{\frac{1}{2}}\tilde{Q}$, $\tilde{Q}$ being ſuch a function of x, that its value is ± 2 when x is $= 0$. Whence ($x^{\frac{1}{2}}\tilde{Q}$ being imaginary when x is negative,) it appears, that the curve is not continued on the left of the point A. Therefore, by what is ſaid above, there muſt be a cuſpid of the ſecond kind, at A, as in Fig. 52.

And by purſuing the ſame method, we may be ſatisfied in other ſuch ambiguous caſes.

Conſe-

Consequently, the value of q being imaginary, the value of $\overline{-1}|^{m} \times q$ is real: and, by Corol. 6. of the last Art. the ordinate BD ($= d$) touches a continued arch, when (AB) the correspondent abscissa is equal to b; and the arch so touched is on the left of the said ordinate.

EXAMPLE III. *Suppose* $x^3 - 3xy^2 + y^5 = 0$.

Then will $[x \perp y]$ be $= \frac{3x^2 - 3y^2}{6xy - 5y^4}$.

Supposing the numerator $3x^2 - 3y^2 = o$, we, from thence and the other equation, find $x = o$, or $x = \sqrt{2}$, or $x = -\sqrt{2}$; and the respective values of y are o, $\sqrt{2}$, and $-\sqrt{2}$.

Supposing the denominator $6xy - 5y^4 = o$, we, from hence and the first equation, find $x = o$, or $x = \frac{9}{5^{\frac{5}{4}}} \times \sqrt{2}$, or $x = -\frac{9}{5^{\frac{5}{4}}} \times \sqrt{2}$, and the respective values of y are o, $\frac{3}{5^{\frac{3}{4}}} \times \sqrt{2}$, and $-\frac{3}{5^{\frac{3}{4}}} \times \sqrt{2}$.

1st. Taking b equal to o, d will be $= o$, and $d + \overline{x - b}|^{m+1} \times Q = Qx^{m+1}$. Therefore $x^3 - 3Q^2x^{2m+3} + Q^5x^{5m+5}$ is $= o$; where, it is obvious, m may be $= o$, also $m = -\frac{2}{3}$.

If m be equal to o, $1 - 3q^2$ will be $= o$: Hence q is found $= \overline{\frac{1}{3}}|^{\frac{1}{2}}$, or $q = -\overline{\frac{1}{3}}|^{\frac{1}{2}}$. By these two values of q it appears, that two branches of the curve intersect each other at A, the Fig. 53.
point where x begins: and that the tangents (AG, Ag) to those branches, at A, make equal angles (GAH, gAb) with the base AE; and such that GH and gb (which are supposed parallel to the ordinates of the curve) are to AH and Ab respectively, as $\overline{\frac{1}{3}}|^{\frac{1}{2}}$ to 1.

If m be equal to $-\frac{2}{3}$, $-3q^2 + q^5$ will be $= o$: Hence we have $q = 3^{\frac{1}{3}}$. Therefore, by Corollary 4. of the last Article,

A is a point of contrary flexure, in another branch of the curve; the tangent to which, at that point, is parallel to the ordinates; and, q being positive, the value of y corresponding to this branch will be positive when x is positive.

2dly. Taking b equal to $\pm\sqrt{2}$, d will be $=\pm\sqrt{2}$. Therefore $\frac{R}{\overline{x-b}|^m}$ (abridged according to our Remark) is $=$

$$\frac{3}{\overline{x-b}|^m}\times\frac{x^4-2}{6bd-5d^4}=\frac{3}{\overline{x\mp\sqrt{2}}|^m}\times\frac{\overline{x-\sqrt{2}}\times\overline{x+\sqrt{2}}}{6bd-5d^4}=\overline{m+1}\times q$$

when x is $=b$. Whence it is evident, that m is $=1$, and $q=\frac{3b}{6bd-5d^4}$. Therefore, q being negative when b is $=\sqrt{2}$ and positive when b is $=-\sqrt{2}$, it follows, from what is said above, that the ordinate BP ($=\sqrt{2}$) is a *maximum* when (AB) the correspondent abscissa is $=\sqrt{2}$: and that the ordinate bp ($=\sqrt{2}$) is a *minimum* when (Ab) the abscissa corresponding thereto is equal to $\sqrt{2}$; b being on the contrary side of A, from B; and the point p below the base.

3dly. Taking b equal to $\pm\frac{9}{5^{\frac{1}{4}}}\times\sqrt{2}$, d will be $=\pm\frac{3}{5^{\frac{1}{4}}}\times\sqrt{2}$. Therefore $\frac{R}{\overline{x-b}|^m}$ (abridged agreeable to our Remark) is $=$

$$\overline{x-b}|^{-m}\times\frac{3b^2-3d^2}{-15d^3q\cdot\overline{x-b}|^{m+1}}=\overline{m+1}\times q \text{ when } x \text{ is } = b.$$

Whence it is manifest, that m is $=-\frac{1}{2}$, and $q=\overline{\frac{2b^2-2d^2}{-5d^3}}\Big|^{\frac{1}{2}}$. Consequently, the value of q being imaginary or real, and the value of $\overline{-1}|^m\times q$ real or imaginary, according as b is taken equal to $\frac{9}{5^{\frac{1}{4}}}\times\sqrt{2}$ or $-\frac{9}{5^{\frac{1}{4}}}\times\sqrt{2}$; the ordinate EF (by Cor. 6. of the last Art.) touches a continued arch, when (AE) the abscissa answering thereto is equal to $\frac{9}{5^{\frac{1}{4}}}\times\sqrt{2}$, and the arch so touched is on the left of the said ordinate: likewise the ordinate ef touches a continued arch, when (Ae) the correspondent abscissa

is

is equal to AE; e being on the contrary side of A, from E; and the point f below the base; and the arch so touched is on the right of the said ordinate ef.

6.

When m is $= o$, and the curve is continued on both sides of the ordinate d, that ordinate may pass through a point of contrary flexure, the tangent to which is oblique to the base and ordinate.—Suppose P to be such a point in the curve Pf, and let Pg be a tangent at that point. Then, the ordinate ef, intersecting that tangent in g, being called y; the ordinate BP, d; and the abscissas AB and Ae, b and x respectively; $ef - eg$ will be $= y - d - \overline{x-b} \times q$, q denoting the value of $[x \perp y]$ when x is $= b$, i.e. q is $= d \div$ Subtang. Which value of $ef - eg$, it is manifest, will (as when it relates to Fig. 54.) be positive when x is greater than b, and negative when x is less than b; or (as when it relates to Fig. 55.) positive when x is less than b, and negative when x is greater than b. Therefore, by Chap. 4. Art. 3. $\overline{x-b}|^{n} \acute{Q}$ may be assumed $= y - d - \overline{x-b} \times q$; n being an odd number, or a fraction whose numerator and denominator are both odd numbers; and $\acute{Q}$ such a function of x, that neither it nor its reciprocal shall vanish upon taking x equal to b. From which assumed equation we have $y = d + \overline{x-b}.q + \overline{x-b}|^{n}.\acute{Q}$. Therefore, y being before supposed $= d + \overline{x-b}.Q$, it is evident $q + \overline{x-b}|^{n-1} \times \acute{Q}$ will be $= Q$; where, it is plain, n must be greater than 1:

Fig. 54. 55.

Let therefore $q + \overline{x-b}|^{n-1} \times \acute{Q}$ be substituted instead of Q, in any equation found by substituting, and taking m equal to o, as in the preceding article: And then, supposing $\ddot{q}$ to be the value of $\acute{Q}$ when x is $= b$, n and $\ddot{q}$ will be determined in the same manner as are m and q in that article.

By the value of n (so determined) it will appear whether the ordinate d corresponds to a point of contrary flexure, or not; and,

and, if it does correspond to such a point, the value of $\ddot{q}$ will shew the tendency of the curve on both sides thereof, the convexity of the branch on the right of that ordinate being downwards or upwards, according as $\ddot{q}$ is positive or negative.

EXAMPLE. *In Examp.* 3. *of the last article,* y *is* $= 0$, *when* x *is* $= 0$: and by substituting Qx^{m+1} instead of y, in the equation of the curve, we get $x^3 - 3Q^2x^{2m+3} + Q^5x^{5m+5} = 0$; from whence, by taking m equal to 0, and dividing by x^3, we have $1 - 3Q^2 + Q^5x^2 = 0$. Hence q (the value of Q when x is $= 0$) is found $= + \sqrt{\frac{1}{3}}$.

Substituting $q + x^{n-1}\ddot{Q}$ instead of Q, we have

$$\left.\begin{array}{l} 1 - 3q^2 - 6qx^{n-1}\ddot{Q} - 3x^{2n-2}\ddot{Q}^2 \\ + q^5x^2 + 5q^4x^{n+1}\ddot{Q} + 10q^3x^{2n}\ddot{Q}^2 + 10q^2x^{3n-1}\ddot{Q}^3 + \\ \quad 5qx^{4n-2}\ddot{Q}^4 + x^{5n-3}\ddot{Q}^5 \end{array}\right\} = 0.$$

Here, it is obvious, n must be $= 3$; and consequently $q^5 - 6q\ddot{q} = 0$.

Therefore $\ddot{q}$ is $= \frac{q^4}{6}$, $(= \frac{1}{54}$, as well when q is $= -\sqrt{\frac{1}{3}}$, as when q is $= +\sqrt{\frac{1}{3}}$;) and it appears that the two branches of

Fig. 53. the curve, whose tangents at the point A are AG and Ag, have each a point of contrary flexure at A, as in Fig. 53.

7.

Fig. 54. 55. P being a point of contrary flexure, and the abscissa and ordinate corresponding thereto being called β and δ respectively; and the value of $[x \perp y]$, when x is $= \beta$, being denoted by q; $y - \delta - \overline{x - \beta} . q$, as is shewn in the preceding article, will be positive when x is greater than β, and negative when x is less than β; or positive when x is less than β, and negative when x is is greater than β, how near ſoever x be taken to β. Therefore, $\frac{y-\delta-\overline{x-\beta}.q}{\overline{x-\beta}|^2}$ being poſitive or negative according as $y-\delta-\overline{x-\beta}.q$ is poſitive or negative, it follows from Art. 1. Chap. 4. that the value of $\frac{y-\delta-\overline{x-\beta}.q}{\overline{x-\beta}|^2}$, or its reciprocal, will be $= o$, when x is $= \beta$.

Now, by what is taught in Chap. 2. the value of $\frac{y-\delta-\overline{x-\beta}.q}{\overline{x-\beta}|^2}$, when x is $= \beta$, (i. e. when y is $= \delta$,) is $=$ the value of $\frac{[x \ddot{\perp} y]}{2}$.

Conſequently $[x \ddot{\perp} y]$ is $= o$, or $\frac{1}{[x \ddot{\perp} y]} = o$, when the ordinate y correſponds to a point of contrary flexure: And the ſame concluſion follows, from conſidering that $[x \ddot{\perp} y]$, the quotient of the ordinate divided by the ſubtangent, is then a *minimum*, as in Fig. 54. or a *maximum*, as in Fig. 55.

Moreover, ſuppoſing β to be a value of x, and δ the correſpondent value of y, in either of the equations $[x \ddot{\perp} y] = o$, or $\frac{1}{[x \ddot{\perp} y]} = o$; and aſſuming $\overline{x-\beta}|^n \times \ddot{Q} = \frac{y-\delta-\overline{x-\beta}.q}{\overline{x-\beta}|^2}$, $\ddot{Q}$ being ſuch a function of x, that neither $\ddot{q}$, the value of $\ddot{Q}$ when x is $= \beta$, nor the reciprocal of that value ſhall vaniſh; we have $\frac{y-\delta-\overline{x-\beta}\times q}{\overline{x-\beta}|^{n+2}} = \ddot{Q}$; from whence, when x is $= \beta$, $\frac{[x \ddot{\perp} y]-q}{n+2.\overline{x-\beta}|^{n+1}} = \frac{[x \ddot{\perp} y]}{n+1.n+2.\overline{x-\beta}|^n} = \ddot{q}$. And it follows, from what is before ſaid, in this and the 4th Chap. that, the curve being continued on both ſides of the ordinate δ, if n be an odd number, or a fraction whoſe numerator and denominator are both odd numbers, $[x \ddot{\perp} y]$ (when x is $= \beta$) will be a *minimum* or a *maximum*, according as $\ddot{q}$ is poſitive or negative; and the ſaid ordinate δ ſhall paſs through a point of contrary flexure. But if n be an even number, or a fraction whoſe numerator is even and denominator odd, ſuch ordinate will correſpond to a point of the curve where the curvature is

nothing

nothing or *infinite*, and the concavity of the curve is turned the ſame way on both ſides of that point.

When $[x \dot{-} y]$ is a function of x without y being concerned therein, n and $\ddot{q}$ will be determined in like manner as are m and q when $[x \dot{-} y]$ is ſuch a function of x.

To find n and $\ddot{q}$ when the value of $[x \ddot{-} y]$ conſiſts of terms in which both x and y are concerned, let $\delta + \overline{x - \beta} \cdot q + \overline{x - \beta}|^{n+2} \times Q$ be ſubſtituted for y, in the equation of the curve; or $\delta + \overline{x - \beta} \cdot q + \overline{x - \beta}|^{n+2} \times \ddot{q}$ for y in either of the equations $\frac{[x \dot{-} y] - q}{\overline{n+2} \cdot \overline{x - \beta}|^{n+1}} = \ddot{q}$ when x is $= \beta$, or $\frac{[x \ddot{-} y]}{\overline{n+1} \cdot \overline{n+2} \cdot \overline{x - \beta}|^{n}} = \ddot{q}$ when x is $= \beta$. From either of which equations, after ſubſtitution, n and $\ddot{q}$ may be determined in the ſame manner as m and q are determined above.

Corollary I. n being an odd number, or a fraction whoſe numerator and denominator are both odd numbers, as many ſingle different real values as $\ddot{q}$ has, ſo many different branches of the curve are continued on both ſides of the ordinate δ, each having a point of contrary flexure at P, the point to which the ſaid ordinate correſponds: And two equal values of $\ddot{q}$ denote two ſuch branches, both turned alike; or a cuſpid of the ſecond kind, at that point.

Moreover, (x being ſuppoſed poſitive when the abſciſſa is on the right of the point where it begins) the branch of the curve on the right of the ordinate δ will have its convexity downwards or upwards, according as $\ddot{q}$ is poſitive or negative.

Corollary

Corollary II. n being an even number, or a fraction whoſe numerator is even and denominator odd, as many ſingle different real values as $\ddot{q}$ has, ſo many different branches of the curve are continued on both ſides of the ordinate δ, touching each other in P, and each having its curvature *nothing* or *infinite*, at that point, according as n is poſitive or negative: And two equal values of $\ddot{q}$ denote two ſuch continued arches, both turned one way; or a cuſpid of the ſecond kind, at P.

Moreover, according as $\ddot{q}$ is poſitive or negative, the branch to which it relates will, at P, have its convexity downwards or upwards.

Corollary III. n being a fraction whoſe numerator is odd and denominator even, as many ſingle different poſitive real values as $\ddot{q}$ and $\overline{-1}|^{n} \times \ddot{q}$ have, ſo many cuſpids of the firſt kind will be formed at P: and two equal poſitive real values of $\ddot{q}$, or of $\overline{-1}|^{n} \times \ddot{q}$, denote two ſuch cuſpids, at that point, both pointing one way; or a cuſpid of the ſecond kind there.

Moreover, (x being ſuppoſed poſitive when the abſciſſa is on the right of the point where it begins,) according as the value of $\ddot{q}$ or $\overline{-1}|^{n} \times \ddot{q}$ is real, the branch or branches forming the cuſpid to which ſuch value relates will be on the right or left of the ordinate δ.

Corollary IV. n being $= 0$, as many ſingle different real values as $\ddot{q}$ has, ſo many different branches of the curve are continued on both ſides of the ordinate δ, touching each other in P: and two equal values of $\ddot{q}$ denote two continued arches of equal curvature at P, and both turned one way; or a cuſpid of the ſecond kind, at that point.

Q The

The tangent, at P, to ſuch cuſpid, or continued arches, will be oblique to the baſe and ordinate; and, according as $\ddot{q}$ is poſitive or negative, the branch to which it relates will, at P, have its convexity downwards or upwards.

Having ſaid ſo much concerning curves referred to a baſe, *I think it unneceſſary to add any thing here relating to* SPIRALS;—*to which kind of curves the intelligent Reader will readily apply the above method of reaſoning.*

THE

RESIDUAL ANALYSIS.

CHAP. IX.

Of the ASYMPTOTES *of* curve Lines.

AS a knowledge of the tendency of the infinite branches of a Curve, when ſuch there are, is requiſite for the obtaining a clear idea of the figure of ſuch Curve, and may ſometimes facilitate the buſineſs of finding the moſt diſtinguiſhed points therein; I ſhall therefore point out an eaſy method of finding the Aſymptotes to ſuch branches, by which means their tendency may be very readily diſcovered.

A branch of a curve when infinitely continued may be conſidered as coinciding with its aſymptote. If therefore m and n be reſpectively put for the ſines of the angles which a rectilinear aſymptote makes with any ordinate and the baſe; or for the ſines of the angles which the tangent to a curvilinear aſymptote, when extended infinitely, makes with any ordinate and baſe; m will be to n, as $[v \perp x]$ to $[v \perp y]$, when the curve is infinitely continued. Conſequently, if from the equation of the propoſed curve, (which ſuppoſe clear of ſurds,) we, by reſidual diviſion, ($v - \dot{v}$ being the diviſor,) deduce a ſecond equation, and therein write m and n inſtead of $[v \perp x]$ and $[v \perp y]$ reſpectively, we

 ſhall

ſhall obtain the equation of an aſymptotic line, one dimenſion lower than the given equation, where m and n will be ſome invariable quantities, which will be determined in conſequence of this proceſs. And, the propoſed curve being above the ſecond order of lines, we, from the equation of that firſt aſymptotic line, by the like reſidual diviſion, and again ſubſtituting m and n inſtead of $[v \perp x]$ and $[v \perp y]$ reſpectively, may deduce the equation of a ſecond aſymptotic line, two dimenſions lower than the given equation. Likewiſe, the propoſed curve being above the third order of lines, we, by our ſaid diviſion and ſubſtituting as before, may derive, from the equation of ſuch ſecond aſymptotic line, the equation of a third aſymptotic line, three dimenſions lower than the given equation. And ſo, with great facility, we may proceed with any given equation, till we get equations of aſymptotic lines of every dimenſion lower than the given equation: alſo, from the loweſt of thoſe equations, (viz. that of one dimenſion,) we may, by continuing the ſame proceſs, obtain an equation containing no other unknown quantities but m and n; from whence $\frac{m}{n}$, or $\frac{n}{m}$ will be determined; and then all that is requiſite, in the ſeveral equations of the aſymptotic lines before found, will be known; and conſequently the tendency of ſuch lines, and of ſuch branches of the propoſed curve to which they are aſymptotes.

Scholium I. The poſition of the ultimate tangent * to a parabola being not determinable, ſuch tangent only coming, at an unlimited diſtance, almoſt to a paralleliſm with ſome certain right line: when the aſymptotic line (expreſſed by any equation found as above) is of the parabolic kind, it will appear by ſome equation in the above proceſs, that m (or n) cannot be taken abſolutely equal to its apparent value in the final equation in the ſaid proceſs, without occaſioning an abſurdity; and the impoſſibility of a rectilinear aſymptote will be evident. Therefore, in ſuch caſe, we are to conſider m (or n) only as approaching very near to ſuch apparent value; and accordingly, inſtead of $\frac{m}{n}$ (or $\frac{n}{m}$) ſubſtitute its value obtained from ſome equation before found.

* By *ultimate tangent*, I mean the tangent to an infinitely diſtant point of the curve.

Scho-

SCHOLIUM II. It may happen that the aſymptotic equation of two (or more) dimenſions, ſound by our proceſs, will expreſs two (or more) right lines, inſtead of expreſſing a curve. In which caſe, to find the moſt ſimple hyperbolic aſymptote of the branches to which a rectilinear one ſo ſound relates, ſome farther enquiry is requiſite. Suppoſe the equation of a rectilinear aſymptote (found by the above proceſs) to be $y = Ax + B$. Then, it is obvious, the branches to which ſuch right line is an aſymptote may be expreſſed by the equation $y = Ax + B + x^{p}Q$; p being ſome negative number or fraction, and Q ſuch a function of x, that neither (q) the limit to which it converges, when x (being a very large quantity) is taken greater and greater, nor $\left(\frac{1}{q}\right)$ the reciprocal of that limit, ſhall be $= 0$. And it is likewiſe obvious, that $y = Ax + B + qx^{p}$ is the equation of the moſt ſimple curve which may be an aſymptote to ſuch branches. Therefore let $Ax + B + x^{p}Q$ be ſubſtituted for y in the given equation; and take p of ſuch a negative value in the reſulting equation, that the ſame, after it is divided by the higheſt power of x therein concerned, ſhall not, upon ſuppoſing x infinitely great, conſiſt of leſs than two finite members: From whence q may be determined.

EXAMPLE I. *Let the equation of the curve be* $y^4 - ax^2y^2 + bx^3 = 0$; a *and* b *being invariable poſitive quantities.*

Then, by proceeding as above mentioned, we find the

$$\text{1ſt } \textit{aſympt. equat. } 4ny^3 \begin{matrix} -2amxy^2 \\ -2anx^2y \end{matrix} + 3bmx^2 = 0;$$

$$\text{2d } \textit{aſympt. equat. } 6n^2y^2 \begin{matrix} -am^2y^2 \\ -4amnxy \\ -an^2x^2 \end{matrix} + 3bm^2x = 0;$$

$$\text{3d } \textit{aſympt. equat. } 4n^3y \begin{matrix} -2am^2ny \\ -2amn^2x \end{matrix} + bm^3 = 0:$$

$$\text{alſo } n^4 - am^2n^2 = 0;$$

$$\text{where } \frac{n}{m} \text{ is } = 0, \text{ or } \frac{n}{m} = \pm\sqrt{a}.$$

Taking $\frac{n}{m}$ equal to $+ a^{\frac{1}{2}}$ and $- a^{\frac{1}{2}}$ successively, it appears, by the 3d asympt. equat. that the curve has two rectilinear asymptotes, whose equations are $2a^{\frac{1}{2}}y - 2a^2x + b = 0$, and $- 2a^{\frac{1}{2}}y - 2a^2x + b = 0$; and, by the 2d asympt. equat. that the two hyperbolas, whose equations are $5ay^2 - 4a^{\frac{1}{2}}xy - a^2x^2 + 3b = 0$, and $5ay^2 + 4a^{\frac{1}{2}}xy - a^2x^2 + 3b = 0$, are asymptotes of the proposed curve; also, by the 1st asympt. equat. that the curve has for asymptotes two lines of the third order, whose equations are $4a^{\frac{1}{2}}y^3 - 2axy^2 - 2a^{\frac{1}{2}}x^2y + 3bx^2 = 0$, and $- 4a^{\frac{1}{2}}y^3 - 2axy^2 + 2a^{\frac{1}{2}}x^2y + 3bx^2 = 0$.

Taking $\frac{n}{m}$ equal to 0, we have, from the 3d asympt. equat. $b = 0$: which is absurd, (as b is supposed of some value,) and therefore, according to Schol. 1. shews that there is no rectilinear asymptote parallel to the base; (the final equation $\frac{n}{m} = 0$ only indicating, that the ultimate tangent of some parabolic asymptote has nearly that direction;) and that, for the said asymptotic equation to be a true one, $\frac{n}{m}$ must not be considered as absolutely $= 0$, but only as indefinitely small, so that being multiplied by a very great quantity, the product may be something considerable (viz. $= b$). Therefore, from that consideration, retaining the terms wherein the root $\frac{n}{m}$ is found, (after dividing by m^2,) as being most considerable; and rejecting those in which the higher powers of $\frac{n}{m}$ are concerned, as inconsiderable; we have $\frac{2any}{m} = b$, or $\frac{n}{m} = \frac{b}{2ay}$. Which last quantity being substituted for $\frac{n}{m}$ in the second asymptotic equation, after dividing by m^2 and rejecting the terms wherein $\frac{n^2}{m^2}$ is concerned, we get $bx - ay^2 = 0$. Consequently the conical parabola expressed by this last equation, is an asymptote to the proposed curve. Moreover, by substituting $\frac{b}{2ay}$ instead of $\frac{n}{m}$ in the 1st asympt. equat. it appears that a

line

line of the third order, whoſe equation is $by^3 - a^2xy^2 + abx^3 = 0$, is an aſymptote to the ſame curve.

EXAMPLE II. *Let the equation of the curve be*

$$xy^2 + ey - ax^3 - bx^2 - cx - d = 0:$$

Which is the chief equation in Sir ISAAC NEWTON'S *Enumerat. Linear. Tert. Ordinis.*

Then, by reſidual diviſion, we find the

1ſt *aſympt. equat.* $\begin{matrix} my^2 \\ + 2nxy \end{matrix} + en - 3amx^2 - 2bmx - cm = 0;$

2d *aſympt. equat.* $\begin{matrix} 2mny \\ + n^2x \end{matrix} - 3am^2x - bm^2 = 0:$

$$\text{alſo } mn^2 - am^3 = 0;$$

$$\text{where } \frac{m}{n} \text{ is } = 0, \text{ and } \frac{m}{n} = \pm \frac{1}{a^{\frac{1}{2}}}.$$

It is evident therefore, that, when a is a poſitive quantity, the curve has three rectilinear aſymptotes; one of which, as appears by the ſecond aſymptotic equation, is the principal ordinate, and the other two are expreſſed by the equations $2a^{\frac{1}{2}}y - 2ax - b = 0$, and $2a^{\frac{1}{2}}y + 2ax + b = 0$. Moreover, by taking $\frac{m}{n}$ equal to 0, $\frac{1}{a^{\frac{1}{2}}}$, and $-\frac{1}{a^{\frac{1}{2}}}$ ſucceſſively, in the 1ſt aſympt. equat. it appears that the curve has for aſymptotes the three hyperbolas, whoſe equations are $2xy + e = 0$, $y^2 + 2a^{\frac{1}{2}}xy + a^{\frac{1}{2}}e - 3ax^2 - 2bx - c = 0$, and $y^2 - 2a^{\frac{1}{2}}xy - a^{\frac{1}{2}}e - 3ax^2 - 2bx - c = 0$.

When a is $= 0$, mn^2 is $= 0$; whence $\frac{m}{n} = 0$, or $\frac{n}{m} = 0$. Therefore, if b be not alſo $= 0$, it is manifeſt, from the ſecond aſympt. equat. that the curve cannot have a rectilinear aſymptote parallel to the baſe: But (rejecting $\frac{n^2}{m^2}x$ as inconſiderable) we have,

have, from that asymptotic equation, $\frac{n}{m} = \frac{b}{2y}$. Which last quantity being substituted for $\frac{n}{m}$, in the 1st asympt. equat. after rejecting $\frac{cn}{m}$ as inconsiderable, we get $y^2 - bx - c = 0$. Consequently the parabola expressed by this last equation is an asymptote of the proposed curve.

If both a and b be equal to 0, the 2d asympt. equat. entirely vanishes when $\frac{n}{m}$ is taken equal to 0; and from the 1st asympt. equat. we then have $y^2 - c = 0$, which is an equation to two right lines parallel to the base. From whence we have $y = c^{\frac{1}{2}}$, and $y = -c^{\frac{1}{2}}$. Therefore, in this case, the curve has two rectilinear asymptotes parallel to the base, one above and the other below; and the distance of each from the base is $c^{\frac{1}{2}}$. And it is plain, these asymptotes, when c is $= 0$, both coincide with the base.

It is observable, that, in every case, the principal ordinate is an asymptote.

THE

RESIDUAL ANALYSIS.

CHAP. X.

Of the Diameters *and* Centers *of* curve Lines.

In finding the forms and properties of curves, we frequently have occasion to enquire concerning their diameters and centers: It may therefore be worth while to shew the use of our doctrine in such enquiries.

I.

When any number of parallel right lines, drawn between two branches of a curve, and terminated by those branches, are all bisected by some line, I call such bisecting line a *diameter* of the *first kind*. And, any number of parallel right lines being drawn between two extreme branches of a curve, and cutting the intermediate branches; if those parallels be intersected by some line, (which is not such a diameter as just now mentioned,) so that the aggregate of the parts of any parallel, lying on one side of that intersecting line, and terminated by it and a branch of the curve, be equal to the part, or aggregate of the parts, of the same parallel, terminated in the same manner, lying on the contrary side of the same intersecting line; I then, for distinction sake, call such intersecting line a *diameter* of the *second kind*. When

R all

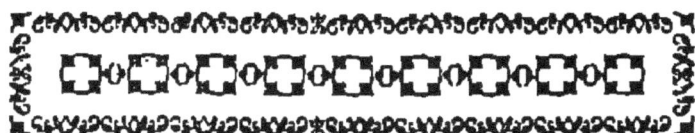

all the rectilinear diameters of this ſecond kind, appertaining to any curve, interſect each other in one and the ſame point; ſuch point of interſection, I preſume, may, not improperly, be called a *diametric pole*.

2.

A point biſecting all right lines drawn through it, and terminated by two branches of a curve, I call a *center* of the *firſt kind*. And, any number of right lines being drawn through a certain point, terminating at the extreme branches of a curve, and cutting the intermediate branches; if the aggregate of the parts of any ſuch line, lying on one ſide of that point, and terminated by it and a branch of the curve, be equal to the part, or aggregate of the parts, of the ſame line, terminated in the ſame manner, lying on the contrary ſide of the ſame point; we may denominate ſuch point a *center* of the *ſecond kind*.

3.

Fig. 56. EFG being any curve, whoſe abſciſſa AB, is called x; and ordinate BF, y: ſuppoſe *efg* to be another line, whoſe abſciſſa Ab, and ordinate *bf*, (parallel to BF,) are denoted by v and w reſpectively: alſo ſuppoſe the right line F*f*, which we will call u, always to make invariable angles with thoſe ordinates. Then, drawing *fb* parallel to AB, and calling the ſines of the angles *fb*F, *f*F*b*, and F*fb*, k, m, and n reſpectively; *fb* will be $= \frac{mu}{k}$, and F*b* $= \frac{nu}{k}$. Conſequently, writing z inſtead of $\frac{u}{k}$, x will be $= v + mz$, and $y = w + nz$. Which values of x and y being reſpectively ſubſtituted for their equals, in the equation of the curve EFG, the nature of that curve will be expreſſed by an equation ſhewing the relation of the three quantities v, w, and z; which may all vary together, whilſt m and n remain invariable. And by expreſſing the nature of the curve in ſuch manner, many uſeful concluſions relating thereto may be very eaſily inferred.

4. Suppoſe

4.

Suppose the equation of the curve EFG to be

$$ax^py^q + bx^ry^s \text{ \&c.} = 0:$$

and suppose that, when $v + mz$ and $w + nz$ are therein respectively substituted for x and y, the resulting expression, viz.

$$a \, . \, \overline{v+mz}|^p \, . \, \overline{w+nz}|^q + b \, . \, \overline{v+mz}|^r \, . \, \overline{w+nz}|^s \text{ \&c. is}$$
$$= A + Bz + Cz^2 + Dz^3 \ldots\ldots\ldots\ldots\ldots Pz^{p+q};$$

$p + q$ denoting the order of the line EFG, and A, B, C, &c. being functions of v and w, in which z is not concerned. Then the value of $[t \perp a \, . \, \overline{v+mz}|^p \, . \, \overline{w+nz}|^q] + [t \perp b \, . \, \overline{v+mz}|^r \, . \, \overline{w+nz}|^s]$ &c. when only z is considered as variable, and unity is wrote instead of $[t \perp z]$, being the same as when v and w are considered as variable, and z as invariable, and m and n are wrote instead of $[t \perp v]$ and $[t \perp w]$ respectively; A, B, C, &c. will be found as follows.

1st. From the above assumed equation, we, by taking z equal to 0, get $av^pw^q + bv^rw^s$ &c. $=$ A.

2dly. By residual division, making $t - \dot{t}$ the divisor, and, on one side of the assumed equation, considering v and w as variable whilst z is invariable, and writing m and n instead of $[t \perp v]$ and $[t \perp w]$ respectively; and, on the other side, considering z only as variable whilst A, B, C, &c. are invariable, and writing unity for $[t \perp z]$; we get

$$amp \, . \, \overline{v+mz}|^{p-1} \, . \, \overline{w+nz}|^q + anq \, . \, \overline{v+mz}|^p \, . \, \overline{w+nz}|^{q-1}$$
$$+ bmr \, . \, \overline{v+mz}|^{r-1} \, . \, \overline{w+nz}|^s + bns \, . \, \overline{v+mz}|^r \, . \, \overline{w+nz}|^{s-1} \text{ \&c.}$$
$$= B + 2Cz + 3Dz^2 + 4Ez^3 \text{ \&c.}$$

Hence, by taking z equal to 0, we have

$$ampv^{p-1}w^q + anqv^pw^{q-1} + bmrv^{r-1}w^s + bnsv^rw^{s-1} \text{ \&c.} = B.$$

 And

And the values of C, D, E, &c. may be found in the same manner.

It is obvious therefore, that, if, in the equation of the curve EFG, (the terms being all brought to one side,) and the several asymptotic equations deduced therefrom by residual division, (as in the preceding chapter,) v and w be substituted instead of x and y respectively; the several expressions, after such substitution, will be the same as those by which (z^0, z, z^2, z^3, &c.) the several powers of z are respectively multiplied when $v + mz$ and $w + nz$ are substituted for x and y in the equation of the curve EFG, as before mentioned. And, by means of such expressions, we may not only determine the asymptotes of curves; but likewise, with the greatest facility, may find their diameters and centers; and moreover, discover many other remarkable particulars relative to the intersections of lines.

Example. *Let the curve be proposed whose equation is given in Examp. 2. of the last chapter: which equation is*

$$xy^2 + ey - ax^3 - bx^2 - cx - d = 0.$$

Then, transmuting that equation, as above mentioned, we have

$$\left.\begin{array}{r} vw^2 + ew - av^3 - bv^2 - cv - d \\ + \overline{mw^2 + 2nvw + en - 3amv^2 - 2bmv - cm} \times z \\ + \overline{2mnw + n^2v - 3am^2v - bm^2} \times z^2 \\ + \overline{mn^2 - am^3} \times z^3 \end{array}\right\} = 0.$$

Now, considering m and n as invariable whilst v, w, and z vary; if z has but two values, and one of them is always as much negative as the other is positive, the line efg, with respect to EFG, will be a diameter of the first kind: Which will be the case when the coefficients of z^3 and z vanish, and, at the same time, the coefficient of z^2 and the terms in which z is not concerned, do not vanish. Therefore, $\frac{m}{n}$ being as determined by the equation $mn^2 - am^3 = 0$, let the equation of the line efg be the coefficient of z put $= 0$, i. e.

$$mw^2 + 2nvw + en - 3amv^2 - 2bmv - cm = 0;$$

and

and that line will be a diameter of the first kind, with respect to the curve EFG. Moreover, $\frac{m}{n}$ having three real values when a is positive, there will then be three such diameters; which will be expressed by the same equations, by which the hyperbolic asymptotes of the curve EFG are expressed, in the second Example in the preceding Chapter; and, efg being such a diameter, Ff will be parallel to a rectilinear asymptote of the same curve.

The equations expressing those diameters are $2vw + e = 0$, $w^2 + 2a^{\frac{1}{2}}vw + a^{\frac{1}{2}}e - 3av^2 - 2bv - c = 0$, and $w^2 - 2a^{\frac{1}{2}}vw - a^{\frac{1}{2}}e - 3av^2 - 2bv - c = 0$. The first of which, when e is $= 0$, expresses two right lines, one coinciding with the base, and the other with the principal ordinate; whereof, the former is a diameter of the first kind, and the latter an asymptote of the proposed curve. By the second of those three equations, w is $= -a^{\frac{1}{2}}v \pm 2a^{\frac{1}{2}}\sqrt{v^2 + \frac{b}{2a}v + \frac{c - a^{\frac{1}{2}}e}{4a}}$; and, by the third, w is $= a^{\frac{1}{2}}v \pm 2a^{\frac{1}{2}}\sqrt{v^2 + \frac{b}{2a}v + \frac{c + a^{\frac{1}{2}}e}{4a}}$. It appears therefore, that if $\frac{b^2}{4a}$ be $= c - a^{\frac{1}{2}}e$, or $\frac{b^2}{4a} = c + a^{\frac{1}{2}}e$, the second or third of those three diametric equations will accordingly express two right lines; one of which will be a diameter of the first kind, and the other an asymptote of the curve EFG: and that, if e be $= 0$, and $\frac{b^2}{4a} = c$, the second and third of the said diametric equations will each express two right lines; whereof, one will be a diameter, and the other an asymptote, as just now mentioned; so that, in this last case, the proposed curve will have three rectilinear diameters of the first kind.

If a be $= 0$, the second and third diametric equations both become $w^2 - 2bv - c = 0$, an equation to a conical parabola; which differs a little from the parabola that, in the same case, is an asymptote to the curve: and this diametric parabola will bisect lines drawn parallel to the base, and terminated by two branches of the proposed curve.

If

If both a and b be equal to o, the coefficient of z^2 will vanish when $\frac{n}{m}$ is = o; which shews that no right line parallel to the base can cut the curve EFG in more than one point. It is therefore obvious, that, in this case, (as well as when a is negative,) the curve will have but one diameter of the first kind, which will be expressed by the first of the three diametric equations above written.

It is manifest, that, if the equation of the line efg be the coefficient of z^2 put = o, i. e. $2mmw + n^2v - 3am^2v - bm^3 = o$; let $\frac{m}{n}$ be what it will, (provided it be not a root of the equation $\frac{m}{n} - \frac{am^3}{n^3} = o$,) that line will be a diameter of the second kind, with respect to the curve EFG.

And, $\frac{m}{n}$ being of any value whatever, when b is = o, w in the last mentioned diametric equation will always be equal to o when v is equal to o; and consequently the point where v (or x) begins will, in such case, be a diametric pole.

In finding a center of either kind, by our method, efg must not be considered as a line, but as a fixed point; and therefore v and w must be considered as invariable, in the coefficients of the several powers of z, whilst $\frac{m}{n}$ and z vary. Now, as $\frac{m}{n}$ must be variable, it is plain, that $mn^2 - am^3$, the coefficient of z^3, cannot be equal to o. Therefore, for the proposed curve to have a center of the first kind, the coefficient of z^2 must vanish without determining the value of $\frac{m}{n}$; and the terms wherein z is not concerned must also vanish, that the remaining terms in the equation of the curve may be divisible by z. Consequently, when a line of the third order has a center of the first kind, such center will always be some point in that line. Moreover, for $(2mmw + n^2v - 3am^2v - bm^3)$ the coefficient of z^2 to vanish without determining the value of of $\frac{m}{n}$, it is evident b, v, and w muſt each be $= 0$; and ſince, the terms in which z is not concerned being ſuppoſed $= 0$, w cannot vaniſh when v vaniſhes, unleſs the term d be wanting; it follows, that the curve will have no center of the firſt kind, unleſs b and d be each $= 0$; and that, if the terms bx^2 and d be wanting in the propoſed equation, the point where the abſciſſa (x) begins will be a center of that kind.

It is obvious, that, if the coefficient of z^2 vaniſhes without determining the value of $\frac{m}{n}$, and, at the ſame time, the terms in which z is not concerned do not vaniſh; the curve will have a center of the ſecond kind. Which will be the caſe, when b, v, and w are each $= 0$; and the term d, at the ſame time, is not wanting: the point where the abſciſſa (x) begins being then ſuch a center.

This Chapter might be extended to a great length, in expatiating on the uſefulneſs of our method of tranſmuting the equation of a curve; but what is already ſaid may ſuffice to enable the intelligent Reader to make a farther application of it, at his pleaſure.

In the inveſtigation of propoſitions, as well phyſical as geometrical, the *relation between two variable quantities* is frequently the object of our enquiry; and ſuch relation never appears more clearly, than when one of thoſe quantities is conſidered as the *abſciſſa*, and the other as the *correſpondent ordinate* of a curve, and the form of the curve is properly aſcertained.—By conſidering the variable quantities in that light, miſtakes are prevented, and the moſt ſatisfactory concluſions obtained, in various intereſting diſquiſitions, particularly in the reſolution of problems relating to the *maxima* and *minima* of quantities.—Now, in aſcertaining the form of a curve from the equation thereof, the Articles in the laſt three Chapters will, I preſume, be found of conſiderable

www.ingramcontent.com/pod-product-compliance
Lightning Source LLC
Chambersburg PA
CBHW021816190326
41518CB00007B/622
9783742823601